CAN MAN PREVAIL?

Books by Hugh Miller

History and Science

An Historical Introduction to Philosophy

The Community of Man

Progress and Decline—The Group in Evolution

CAN MAN PREVAIL?

HUMAN LIFE IN ITS EVOLUTIONARY ENVIRONMENT

Hugh Miller

An Exposition-University Book

EXPOSITION PRESS HICKSVILLE, NEW YORK

First Edition

Inquiries should be addressed to
Exposition Press, Inc., 900 South Oyster Bay Road,
Hicksville, N.Y. 11801

ISBN 0-682-47981-0

Library of Congress Catalog Card Number: 74-76030

Printed in the United States of America

Contents

Preface

The twelve short chapters which follow offer a condensed and it is hoped more readable summary of a lengthier work first published in 1965 by The Ward Ritchie Press of Los Angeles, entitled **PROGRESS AND DECLINE: The Group in Evolution,** and later taken over and distributed by the Pergamon Press of Oxford, England.

Their initial purpose is to call attention to the important but hitherto neglected reproductive processes, genetic and other, which everywhere work within living populations to perpetuate plant and animal species and to maintain human societies. It is by means of these intragroup processes that individual organisms are bound into integral species and societies, given their powers of self-regeneration, and kept adaptable to external change so that they may long persist; and it is to those same processes, far more than to the process of natural selection that merely destroys what is individually little adapted to its immediate surroundings, that we must go for our understanding of the evolutionary change that has made the surface of this earth what it is today.

Our earth has become a man-dominated planet, one in which accelerant human increase and an explosively expanded industrial economy now threaten the natural resources—even the water and atmosphere—upon which all life must indefinitely depend. Suddenly made aware of this precarious situation, we still think to escape from it by correcting this or that detail of our more flagrantly destructive exploitation of nature; but the ecology of nature is all of a piece, and it is to be saved from destruction only by means of an organized, coordinated, and worldwide management which a divided and increasingly dissolute mankind is powerless to provide.

Our most pressing need is accordingly of an unprecedented advance of moral responsibility, social wisdom, and human power, such as can come only through a forward leap to a new and large comprehension of the true status of man. For this accession of human faith we must look to our knowledge, steadily bettered but as yet unused, of the so-called evolutionary processes which are in truth the very dynamo of nature, inasmuch as it is they, and nothing else, that brought into being and gave shape and character to whatever describably exists—to man, to living nature, to earth and ultimately to cosmos itself. It is finally to this as yet undeveloped science of creation—the only discipline, it would seem, with a prospective future—that these pages are dedicated.

CAN MAN PREVAIL?

I

After Darwin

More than a century has now elapsed since Darwin published **The Origin of Species.** During the remaining decades of the nineteenth century its evolutionary hypothesis gained scientific support and general acceptance. It stimulated interest in the geological description of the rise and erosion of great mountain masses, and in the paleontologist's study of fossil remains preserved in sedimentary rocks. This provided evidence of many extinct forms of life like nothing found on earth today. Darwin's successful establishment of evolutionary science, after the failure of earlier efforts to do so, was as much due to his theory of natural selection as to the abundant evidence with which he justified his epoch-making hypothesis; for that theory presented evolutionary change as a necessary consequence of the selective pressure exerted upon living populations by their physical environment and by other species.

Later, in our own century, this theory was to become the main support of an impressive science of ecology, which found everywhere in living nature the observable evidence of evolutionary change. Ecology was to have many important applications in agriculture, husbandry, horticulture, forestry, and other fields of practical activity; and, as a result of this technological use, there came a gradual shift of both scientific and public interest in respect to evolutionary science. The attention earlier focused upon the paleontologist's reconstruction of evolutionary history was gradually diverted, to be bent instead upon the ecologist's description of evolutionary change as this observably proceeds in populations living on earth today.

In this way evolutionary science was brought into line with other descriptive science. Ever since Bacon in the early seventeenth century issued his call for a knowledge that would serve mankind and revolutionize human economy, natural science had been increasingly pursued as a means to technological progress, rather than as a pursuit of knowledge for its own sake, as we liked to say, meaning for the larger comprehension it brings of the world we live in. That this practical interest did not prevent scientific progress is shown by the remarkably rapid advance of physical and chemical science during the present century; but it did work as a deterrent to the proper development of evolutionary science, and this for two very different reasons.

The first of these is that evolutionary science is properly an historical science concerned with the long past, and comprises no part of what we now designate as theoretical science; and its description of evolutionary history does not materially support technological invention, as does most or all theoretical knowledge, so that it must be pursued wholly "for its own sake" as a means to a better comprehension of our world. A physicist or other theoretical scientist is inclined to accept the term "historical science" as no more than a courtesy title. His own science is advanced by means of hypotheses that can be confirmed by the verifiable predictions they allow of future occurrence; but no reconstruction of past history can be so validated, what yet may be evolved being quite unpredictable. This adverse view seems to be that of most biologists too, for evolutionary history is still regarded as a special perquisite of the paleontologist, and as an interesting offshoot from life science rather than as an integral part of it. Nor do scientists often stop to consider how much their description of the life they observe today is shaped by what has been learned of its genealogical descent.

The second reason is of quite another sort. The use of descriptive science as a means to technological invention has directly worked to obscure from the scientist the most important determinants of evolutionary change. Technological progress is actually a continuous readaptation of man himself to his external environment; and it is accordingly not surprising that attention was increasingly given to the readaptive changes that may better fit plants and animals to their immediate surroundings. It was this exclusive emphasis, of course, that gradually transformed what had been a study of evolutionary history into an ecological science descriptive of the adaptive changes observable in plant and animal populations existing today; and this was a trend that seemed to be required by any acceptance of Darwinian theory itself, this being only an explicit definition of the process of natural selection by which such adaptive change is brought about. And the trend has persisted down to the present time, although first Darwin himself, and following him in this century an increasing number of paleontologists, geneticists, and other life scientists, have worked to widen evolutionary theory, by showing that the continual readaptation of living populations to their external surroundings is only a subsidiary cause, and by no means the major cause, of the long evolution that brought into being what has lived on earth.

Two sciences especially have provided this new evidence, these being paleontology and genetics. By their diligent discovery and increasingly able reconstruction of fossil remains, paleontologists have revealed the existence in past ages of forms of life quite unlike those observable today; and in this way they not only helped to establish the actuality of evolutionary change, but also showed how important and spectacular such change could be. In this century, however, they have moved more ambitiously to a comprehensive reconstruction of evolutionary history, showing how the great families, orders, and classes of plant and animal life have successively expanded and declined. But of even greater importance for our understanding of the causal agencies directing the history is the recent development of population genetics, a discipline which applies the knowledge of genes and genotypes gained by microscopic analysis in the laboratory to living groups observable in nature.[1]

These new approaches require us to broaden our concept of natural selection. This had been taken to cover only the perpetual selection of better adapted individuals, these being able to survive through growth and to transmit their adaptive type to a new generation. There clearly proceeds also a much larger sort of selection, one that in every age carries most species to extinction, while bringing some few species to such expansive multiplication that they more than replace those extinct. This second sort of selection works more slowly but no less rigorously than the first, few of the species existing at any time persisting into the next epoch. That most species should be headed for extinction is a disturbing thought, until one remembers that a few persistent species have always been mightily multiplied, again and again to replenish the earth with plant and animal life.

Why there should be this high mortality of species is not at once apparent. One would expect that the ceaseless elimination of what is poorly adapted to its environment, this leaving only what is well adapted to reproduce its kind in progeny, would be a cause of population increase, and this indeed it is. The continual establishment of what is closely adapted to prevailing conditions has repeatedly resulted in the expansive multiplication of single species into new genera, families, and still larger groups of cognate species. However, a study of the genetic processes that delimit a species in space and also determine its temporal extension makes it equally evident that the progressive establishment of what is uniformly adapted to present conditions leaves living populations steadily less adaptable to environmental change. We are thus shown that it is one and the same group-process which first expansively multiplies species only later to bring them to extinction.

In order to persist, a living group must be continually regenerated, with each short-lived generation producing a

persistent would leave fossil remains still discoverable and identifiable today.

Nevertheless, spotty and incomplete though the fossil record may be, it supports quite a remarkable reconstruction of evolutionary history, and to study this history is to confirm the conclusion that the multiplication of species and the extinction of species are the earlier and later phases of the same evolutionary change, wide expansion usually being the prelude to progressive decline. If the major theme of the great drama of evolution must be found in the sustained advance from some primitive sort of life to human form, the ever repeated sequence of expansion followed by decline may be taken to reiterate its minor theme.

Animal history provides a veritable library of illustrations of this large evolutionary change. The zoologist brings all animal species into about a score of major groups designated as phyla, and the study of the fossils deposited at earlier and later times by these large groups has disclosed a good deal about the phyletic histories. These go back some 400 or 500 million years to the first deposits of identifiable fossils—these dates usually being accepted, probably mistakenly, as the times when the different groups originated. Of the many phyla, one became wholly extinct about 200 million years ago, many others are less widely expanded today than at earlier periods of their history, and only a few appear still to maintain their maximum expansion.

A survey of the phyletic histories shows further that most or all animal groups have undergone expansion and decline more than once during their long persistence. This was undoubtedly due in part to wide fluctuations of the environmental conditions, now more and now less favorable to individual growth, that have been imposed upon living populations since life began; but a closer examination of the group histories informs us that particular species may come earlier or later to the expansive phase that multiplies them into new genera, new families, new orders, and even new classes of cognate species. Species belonging to one and the same phylum may come to wide expansion in very different geological periods, separated by as much as 100 million years, and this, too, would account for the apparent recurrence of expansion followed by decline.

Best known, because longest and most diligently studied, is the animal phylum that contains the vertebrates, the history of this group indicating the course that was taken in the evolutionary progress leading to ourselves. This phylum comprises eight classes of vertebrate and near-vertebrate species. Four of these are aquatic groups, and respectively contain jawless fishes, armored fishes, cartilage fishes, and bony fishes. The remaining classes are of terrestrial vertebrates, and contain amphibians, reptiles, birds, and mammals. The fossil record of this great phylum goes back between 400 and 500 million years, but it does not show the eight vertebrate classes coming down the ages side by side!

The oldest vertebrate fossils are those of jawless fishes. This class, beginning its recorded history some 425 million years ago, came after 100 million years of fairly wide expansion to such severe depletion that it disappears entirely from the fossil record, although the lampreys are still with us to bear witness to its obscure persistence. Not very long after the jawless fishes, geologically speaking, arose the armored fishes, so-called because of their external skeletons; and this class proceeded to rapid and wide expansion only to come to complete extinction 250 million years ago. Next to enter the record were the cartilage fishes, known to us today in the sharks, rays, and their ilk, and this class, after recovering from a setback that followed its initial expansion, came again to decline when the bony fishes arrived at their expansive phase. This last aquatic class may be as old as is the cartilage group, but for about 100 million years it remained little expanded—then belatedly proceeding to the very wide expansion that has made and kept it the dominant aquatic group.

Next to appear were amphibian species. These are usually

included among the terrestrial vertebrates. However, when we consider the watery habitats, the external adaptations, and in particular the modes of reproduction of these frogs, newts, and salamanders, we are more inclined to assign them to the aquatic vertebrates, especially as their history likewise resembles that of the earlier fishes. Their fossil record goes back about as far as that of the bony fishes, and they early proceeded to their maximum expansion, which was well maintained until some 200 million years ago, when the group came to serious and prolonged decline.

Reptilian fossils do not appear until much later, some seventy or more million years after those of their aquatic predecessors. This class at once proceeded to remarkably wide and well diversified expansion, its many species covering the earth and even invading the seas and the air. Although much depleted as were all animal groups during the so-called Permian-Triassic crisis, a period when great mountain-building and glacial climates subjected both aquatic and terrestrial life to unprecedented pressures, the reptiles more than made good their losses and maintained themselves as the dominant terrestrial group until 100 million years ago, when they too came suddenly to catastrophic decline. The great brontosaurs, tyrannosaurs, and ichthyosaurs pass from the scene, and with them go most of the smaller reptiles, to leave existing today, out of the sixteen orders of that once great class, only three orders respectively containing turtles, snakes and lizards, and crocodiles.

The fossil records of the birds and the mammals go back only about 200 million years, and, during the first half of this period, neither class underwent much expansion, their later wide expansion proceeding concurrently with reptilian decline. This raised an interesting question. Did these two classes, the last vertebrate groups to reach their phase of wide expansion, merely replace the reptiles, annexing reptilian habitats now left vacant? Or did they effectively displace the reptiles, somehow bringing most of the reptilian orders to extinction? However this may be, the birds and mammals remained the dominant vertebrate groups, the birds still expanding while there is some evidence of mammalian decline.

The history of the vertebrate phylum, taken in its entirety, must compound the several histories of the eight vertebrate classes, much as a history of Europe compounds the histories of a number of European peoples. When so comprehended, the phyletic history appears to describe a successive replacement of earlier expanded classes by later expanded classes. Much the same sort of replacement appears within the class histories, where later expanded orders replace those earlier expanded. It appears again in the histories of orders, families, and even of genera, wherein extinct species may be replaced by newly established species.

The analysis of the genetic process that is a species, which was presented in the preceding chapter, indicated that species likewise owe their persistence to their replacement of defunct or much depleted local populations. It would accordingly appear that replacement is indeed a universal feature of the evolution, the continual replacement of populations within species being the dynamo that expands single species into new genera of species, these into new families of species, and families into orders, classes, phyla, and even the great kingdoms of plant and animal life.

This sustained expansion was evidently the achievement of a small minority of species, these being able to persist and to undergo repeated expansion in face of the enormous extinction that in every epoch wiped out the great majority of species. Unless we are willing to ascribe this indefinite persistence of some few species to sheer chance, or even more irrationally to the preferential treatment accorded to those species by a discriminating environment, it would appear that we are directed to a better understanding of the third feature of the evolution of life, that of progress, for the explanation of this phenomenon.

III

The Process That Is a Species

The Greek word **species** originally denoted just a kind or sort of thing. Aristotle gave the word a new and special meaning through his doctrine of specific forms. The many kinds or species of things observable on earth, he taught, are really a manifestation of certain divine beings located somewhere in outer space, which project into a chaotic earthly matter their many specific forms. This somewhat fantastic notion was transmitted through two thousand years down to our modern age, when a young Swedish botanist was at last able to bring life science back to earth again.

In 1729, Karl von Linne, known to us as Linnaeus, wrote a short treatise upon sex in plants. By good fortune this was given some scientific attention, which encouraged its writer to begin just a few years later the publication of his celebrated system of classification. Plants were classified by means of a comparison of the observable differences of reproductive development—or, as Linnaeus called them, the different "sexual characteristics"—which are found to separate species one from another. It is regrettable that neither Linnaeus nor his collaborators realized that animal species, too, are properly to be identified by way of their different regenerative processes. The classification was thereafter steadily widened in an effort to include within its many categories all the known species of plant and animal life.

Not until another century had passed did there come a belated but successful inauguration of evolutionary science. This at once gave new and deeper significance to the divisions of the Linnaean classification, for it was now made apparent that the several categories of the system—the **genus** containing several species, the **family** made up of several genera, the **order** comprising a number of families, the **class** composed of orders, and finally the **phylum** made up of classes—must all denote the diverging lines of species descent. In short, the classification of species delivers a foreshortened perspective upon the genealogy of plant and animal life.

This use of the classification implies that the observable characters by means of which species are assigned to their proper genera, families, and wider categories are in and of themselves a reliable indication of their genealogical descent, and in this case we cannot take the distinctive characters which separate species to be the adaptive characters that fit individuals to their external surroundings. If species were to be specifically differentiated by these adaptive characters, which tell us what sort of environment has left its mark upon them, bats would have to be classed with birds, and porpoises and whales would have to be included among the fishes, instead of being brought into the mammalian class to which bats, porpoises, and whales alike belong. But, if it is not by means of their types of adaptive development, then it can be only by way of their specific modes of reproductive development that species are to be classified.

Whatever we choose to call it, we must give due recognition to a collective power of self-regeneration which enables the short-lived generations of a living population to produce out of their own substance a next generation. When a species is multiplied into the several species that establish a new genus, that progenitor species recognizably transmits to the multiple species its own specific mode of self-regeneration, this being what requires us to assign the multiple species to one and the same genus. However, those reproductive modes must be transmitted with a difference, for the multiple species, each of which is fertile enough within itself, are no longer interfertile with one another. It is this intersterility, as Linnaeus correctly perceived, that actually separates the species one from another.

The process of species multiplication may be so long continued that the genus is expanded into a family, an order, a class, and eventually a great phylum. When it is so continued, the reproductive modes of the much multiplied species become ever more widely differentiated, yet never so far as to erase the similarities which bear witness to their common descent. The classification of species is accordingly a classification of the many modes of sexual reproduction that regenerate living populations and that also effectively separate species one from another.

If these seemingly obvious implications of the Linnaean system of classification had been clearly drawn and given explicit statement, the subsequent development of evolutionary science might have been very different and its progress much more rapid than it was. There would also have come a proper recognition of the exceptional genius of Linnaeus, whose rightful place in life science may compare with that of Newton in astronomy; for it was his correct perception of the reproductive characters that actually separate species and regenerate life which first suggested and then directed his construction of the classificatory system, thereafter to be consistently confirmed by a host of scientists who labored to bring the system to completion. A life science without Linnaeus would resemble an astronomy that was ignorant of the force of gravitation.

Unfortunately this opportunity to render credit where credit was due went unseen, and the oversight was really that of the pioneers of evolutionary science themselves. Their first and most desperate need was to meet and if possible appease the vehement opposition that was directed against their enterprise considered so novel at the time. This opposition came as much from conservatively minded scientists as from the general public. Lamarck and other pre-Darwinian advocates of evolution had failed because they presented evolutionary change as the work of vital processes said to be inherent within species themselves; for this teaching was quickly discredited by scientists, ostensibly on the ground that the most painstaking investigation had been unable to identify or locate those "vital powers." The doctrine was accordingly labelled "vitalism" and authoritatively proscribed.

This label stuck, and in Darwin's day anyone who ventured to advance the notion of an evolution of life was still summarily dismissed as just another vitalist. Darwin was therefore most careful, when he came to write **The Origin of Species,** to forestall any possible confusion of his theory with that of the proscribed vitalism. He presented evolutionary change as no more than an inevitable consequence of the individual differentiation of adaptive character which allows some individuals to reach maturity and reproduce their kind, while others perish during growth. Further, his theory of natural selection, or the survival of the fittest, located the agency directly responsible for evolutionary change in the external environment, and not within the species. Here was a doctrine that stayed well clear of the discredited vitalism, and it was one that might prove acceptable to all scientists.

That epoch-making work was first published in 1859. Twelve years later, in 1871, Darwin published a second major treatise entitled **The Descent of Man and Selection in Relation to Sex.** In this book he showed that he had come to see in the earlier principle of natural selection only a beginning and by no means the completion of evolutionary theory. As its title intimates, and as its description of the evolution of reproductive character and sexual behavior that was to lead to man makes very evident, Darwin was by no means unaware of the implications of the Linnaean system. However, the scientists who followed Darwin had little liking for that work. They had made of **The Origin of Species** their bible, and they set up its narrower theory of natural selection as a dogma, one which defined "orthodox Darwinism" and was designed to supplant the traditional dogma of eternally fixed species once and for all.

The guiding principle of modern science has been its reliance upon evidence provided by factual observation, with complete freedom from any confinement imposed by established dogma, and this, of course, makes any requirement of scientific orthodoxy a contradiction in terms. Nevertheless, the great majority of life scientists have since deferred to that "orthodox" Darwinism, which required evolutionary change to be always and only some readaptation to a changing environment. The effect of this restriction was to prevent any use being made of the vast accumulation of evidence that is incorporated into the classification of species. Just because that evidence provides knowledge of likenesses and differences of reproductive development, and says little or nothing about likenesses and differences of adaptive development, no use could be made of it in theoretical construction nor in a scientific description of evolutionary history.

Yet that evidence could not be utterly ignored. There are pairs of closely related bird species that occupy the same habitats and that display no notable differences of adaptive character or behavior, yet that have to be identified as distinct species simply because they do not interbreed. Species are properly to be identified only as breeding groups, each species being fertile within itself but infertile with other species. Those pairs of intersterile species of birds were evidently comprised not too long ago within single species, but no study of their adaptive characters will explain how they came to be separate species. A clue as to how they became separated is provided by certain other bird species, which are so far strung out in space that populations at opposite ends of the species have become reciprocally intersterile, so that their members when brought together no longer interbreed. The ornithologist explains that this is a result of their long separation. However, he still assigns these terminal populations to one and the same species, and the reason he does so is that they are still fertile enough with nearby populations, and through their gene exchange with these neighbors they contribute their local varieties of genotype to the **gene pool,** or reservoir of richly diversified genotype, which through its continual circulation through the species—or **gene flow—** keeps the species at once integral and one, and also widely adaptable to external change.

Mention was earlier made of the process of bisexual reproduction which brings to each new individual a unique combination of parental genotypes, and thereby ensures an individual differentiation of adaptive character; and also described was the preponderant inbreeding within the several populations of a species, which is what enables a local population to establish and retain the variant genotype that keeps it closely fitted to its particular habitat.

These features of the genetic process that is a species must also have their effect upon reproductive character, no less than they do upon adaptive character. Bisexual reproduction ensures that reproductive development, too, will be individually differentiated, and preponderant inbreeding allows the establishment of varietal genotype ensuring that each local population will be sufficiently fertile and fecund within itself.

In other words, two sorts of natural selection, and not one only, are at work within living groups. There is a continual selection of what is best adapted to the external environment, and in particular to the local habitat; but, in addition to this, a very different sort of selection renders the individuals of a local population steadily more uniform with respect to their reproductive character, so that they become more compatible sexually, more fertile with one another, and consequently more fecund. Some individuals produce many progeny, others few or none, and this reproduction differential allows the establishment of genetic variety making a population steadily more fertile and fecund within itself.

With both sorts of selection continually at work, the first acting to leave populations more closely adapted to their local habitats so that more individuals reach maturity, and the second sort making these survivors more fertile and fecund at maturity, a species will exhibit a tendency to population increase. And we have here a further causal explanation of the expansive process that has repeatedly multiplied single species into new genera, families, orders, classes, and even new phyla of cognate species. However, preponderant inbreeding has also certain other very different consequences.

The preponderant inbreeding that leaves local populations progressively more inbred, and consequently more fertile within themselves, renders them at the same time less fertile with other populations of their species. There is necessarily a corresponding diminution of the outbreeding which through its propulsion of locally variant genotypes through a species keeps its populations genetically diversified, so that they may remain adaptable to changing conditions, and also—still more importantly—well integrated into their species. Preponderant inbreeding works as a centrifugal force, directing the local populations along divergent routes of evolutionary change, and this, although it conditions expansive increase, also leaves the population less interfertile and less firmly integrated into their species. Working against this force is the centripetal force generated by the outbreeding among the populations. This keeps the populations genetically diversified, so that while they may bring fewer of each generation to reproductive maturity, they remain widely adaptable and also sufficiently interfertile to maintain effective outbreeding.

As a rule, it will be the peripheral and outlying populations of a species that are prone to expansive increase as the result of excessive inbreeding. With fewer neighbors, they will outbreed less than do the interior populations that have neighbors on all sides of them. The more central populations will always constitute the most firmly integrated, most widely adaptable, least expansive, but most durable core of the species, and this should be the last part of a species to become extinct.

As the peripheral populations spread over new terrain which is here more favorable and there less favorable to individual growth, densely populated areas will come to be separated by sparsely populated areas, and this will give rise to genetic faults across which outbreeding is slight or intermittent and gene flow correspondingly weak or fluctuating. It is along such faults that a distended species will eventually be disrupted into parts that become steadily less interfertile, and that will persist only if they are able to establish their own systems of gene exchange and gene flow. Most of these disrupted parts are doubtless directed to extinction, but, if some of them should persist, they will do so as distinct species establishing a new genus of closely related species.

Individuals are shortlived, but species may long persist. A species that has been disrupted into two or more parts does not thereby cease to exist, and it may still be identified with the part to which came the central core of the disrupted species, this being its strongest, best integrated, most adaptable, and consequently most durable part. If two or more parts persist, they do so as separate species and establish a new genus, and the disrupted species may now be identified as the species progenitor to that genus, persisting alongside of descendant species which it will probably outlive. It follows that genera, families, and still larger groups of cognate species, all of these being a result of continued species multiplication, may often contain their progenitor species, still persisting among the many descendant species. Among the species now existing on earth there must be many that might yet be identified as the species progenitor to their categories, and some of these species must be very old indeed. We need not expect the reptiles that first walked the earth to look too much like anything that walks the earth today. That first reptilian species, in the 250 or more million years that brought its descendants down to the present, may have undergone evolutionary changes that radically changed its appearance, yet it is just the eminent adaptability that allowed such continual readaptation which would account for its continued persistence.

IV

Vertebrate History

Another important feature of the evolution of life is its absolute continuity. From each living individual and every living group there stretches back an unbroken line of descent eventually lost in some dim and remote obscurity. The multitudinous lines of descent are materially continuous, each short-lived generation receiving from its immediate predecessor a genic inheritance that is then again bodily transmitted to a next generation. The genealogy of life is often portrayed as a sort of ragged fan composed of the many divergent lines of species descent, most of the lines being stopped short of the present to indicate extinction.

It has been said that extinction is the usual fate of species. The description of the species process presented here might rather imply that it is the inevitable fate of species. Must not the preponderant inbreeding that allows living populations to regenerate themselves adequately, but that also leaves them less firmly integrated into their species and less adaptable to external change, sooner or later disrupt every species into splinter parts incapable of persistence?

Fortunately, this cannot be the whole story. There must be somewhere another process that brings to species a surer integrity and a wider adaptability. It is estimated that the species existing on earth today outnumber those of any past epoch. Moreover, must not the species now existing be the termini of lines of group descent that were never, in all of the long past, cut short by extinction? Upon the historical warp of expansion and decline there was woven a tapestry that depicts a sustained progress to new modes of regeneration which gave to some groups their lengthy and seemingly indefinite persistence.

Quite evidently, the recently acquired knowledge of genetic processes does not yet deliver a complete description of the evolution of life. This must be one that provides a scientifically acceptable description of evolutionary history. Unfortunately, the advance of evolutionary science has been seriously retarded by two mental prepossessions which still remain to be outgrown. One of these may be likened to a birth trauma, for it came with the "orthodox Darwinism" that was thought necessary to bring the new science of evolution into being. That doctrine required the scientist to describe any and every evolutionary change, however deep-reaching and far-going this might be, as nothing more than some readaptation to external change.

This restriction upon scientific thought has prevented any recognition of the changes of reproductive character which, far more effectively and much more radically than any readaptation to external conditions, have always directed the evolution of plant and animal life. Yet it was only the fact that this evolution was primarily an advance to new modes of reproduction, these giving to living groups their needed powers of self-regeneration, that made possible a scientific classification of species indicative of the genealogy of life. It would be quite impossible to construct a classification of species based upon a comparison of their many different sorts of adaptive character.

The second of these mental prepossessions, astonishingly enough, is a hangover from pre-evolutionary science. It perpetuates habits of thought deeply implanted by centuries of medieval science, which adhered to the Aristotelian doctrine of fixed species—the very doctrine that "orthodox Darwinism" was expected to eradicate once and for all.

That ancient dogma is primarily responsible for the widespread but mistaken notion that when a species undergoes some notable change of form and appearance it thereby becomes a new species, replacing the old species which must now be regarded as extinct. Yet evolutionary science arose to establish the fact that such evolutionary change is actually a condition of species-persistence, and in no way a cause of species-extinction.

But that old dogma is still very much with us. For example, it makes itself apparent when certain recently unearthed skulls are described as being those of the extinct ancestors of man. Is it not logically impossible for any extinct group to be the ancestor of anything living today? An extinct group cannot have any issue, it is the terminus of its particular line of descent.

It is not impossible, of course, that those hominid skulls should actually be those of our prehuman ancestors, but, if they are so, what they testify to is a notable change of cranial form, our own skulls being twice as big. Nothing like those small prehuman skulls is found in populations living today. We might say that their **form** is no longer extant, but we cannot call the prehumans that carried them extinct, if those hominids still persist in ourselves today.

This is doubtless a rather trivial instance. However, that same misconception invalidates what is still the most widely accepted description of vertebrate history. This presents the eight vertebrate classes as having successively advanced the vertebrate progress to man. Each earlier expanded class is said to have contributed its due quota of that progress by first completing its own evolution, and then giving rise to a somewhat more evolved class that would do the same again. One is reminded of a relay race in which each contestant runs his lap and then hands his baton over to the next runner. However, the class of armored fishes is held to have become totally extinct 250 million years ago, and how should that extinct class have given rise to another class more evolved than itself, or for that matter to anything living at all?

Those two prepossessions, each aiding and abetting the other, have defeated every pursuit of an intelligible description of vertebrate history. They require a description that would violate the principles on which is based any classification of species, and this would leave us without any knowledge of the genealogy of life.

Perhaps no contemporary scientist still openly subscribes to that familiar misinterpretation of vertebrate history, yet few scientists have explicitly rejected it, and none has offered an intelligible alternative to it. The origins of the vertebrate classes, it would probably be alleged, are hidden in impenetrable obscurity. As fossil evidence steadily accumulated and came to be better understood, it was found necessary to push ever farther back the dates at which the earlier expanded and less evolved classes were said to have given rise to their later expanded and more evolved successors. Finally it began to look as if each class had produced its successor very early indeed in its own history, not long after its own origin. After all, it could not be until a vertebrate group had proceeded through considerable expansion that it would leave fossil deposits so extensive that they would come through eons of geologic change and still be discoverable today. It is evidently to the successive expansions of the classes, these bringing them into the fossil record, and not to the origins of the classes, that the fossil remains testify.

This opened the possibility that the vertebrate classes might all have originated at about the same time, each class persisting obscurely until it arrived in its turn at its expansive phase. Presumably it is because this possible interpretation raises problems more insuperable than any it might remove that it has received little attention.

It is difficult enough to explain how species of a less evolved class should ever have given rise to a new class more evolved than themselves. The differences that bring vertebrate species into their respective classes are the widest differences to be found within the whole vertebrate phylum, and they cannot have appeared overnight—they must have been tens or perhaps hundreds of million years in the making. How then, is one to describe how a species of jawless fishes, for example, gave rise to a new genus made up not of jawless fishes but of armored fishes? Was that jawless fish species actually never a jawless fish species at all, but an armored fish species travelling incognito? Did it not suspiciously resemble the ugly duckling of Hans Andersen's fairy tale, a duckling that was never a duck at all, but instead a baby swan?

Yet it is even more difficult to explain how there came to be a first genus of vertebrate species so widely differentiated that each was destined to be multiplied into a distinct vertebrate class. That first vertebrate genus must have been composed of four very different sorts of fishes, and along with these came an amphibian species, a reptilian species, a species of birds, and a species of mammals.

Here is indeed something quite inconceivable! The classification of species requires us to believe that the many species of a vertebrate class are of common descent, and also that the eight vertebrate classes are themselves of common descent; but it does not encourage us to believe that the class of jawless fishes should have arisen concurrently with the class of mammals. This supposition the fossil record makes quite impossible. Moreover, the classification requires the establishment of any class to result from the expansion of a family into that class, and this allows of no derivation of any one class from another class.

However, if we reject both of these misinterpretations of the fossil record, we are obliged to seek some new interpretation, and it must be one that allows a description of vertebrate history which does not do violence to the principles upon which is based the classification of species.

This chapter opened with an appreciation of the continuity of life. There are on earth today about 40,000 vertebrate species. From all of these many species there go back unbroken lines of descent that finally converge in a first vertebrate group which was to become progenitor to the whole vertebrate phylum. Just one of those many lines of descent is our own, but this line is to be distinguished from all other genealogical lines, because it also delineates the sustained progress that was to issue in mankind.

Only along that single line of human descent could there have been advanced the vertebrate progress to man, and it is necessary to infer that only by the single regenerate group that was eventually to be expanded into mankind could that progress ever have been achieved. Living groups being what they are—breeding groups kept separate from one another and better or worse preserved by their particular modes of self-regeneration—the vertebrate progress had to be the sustained achievement of just one self-regenerating group. That continuous progress could not be started here and resumed elsewhere. It could not have been advanced first by one group and then by another group. It could not have been achieved piecemeal by each of the vertebrate classes in turn. It had to be the singular achievement of the single living group that persisted quite obscurely until it was just recently brought to world-wide expansion in mankind.

Although the eight vertebrate classes cannot have cooperatively advanced that progress to man, they can and do tell us a good deal about what sort of progress it was, and indicate something of its earlier and later stages. There is no doubt, for example, that the progressive progenitor group was earlier an aquatic group and later became a terrestrial group. This evidence can be used, however, only on the assumption that the progressive group itself gave rise to all of the vertebrate classes, one after the other. In other words, the species that were multiplied into vertebrate classes diverged at earlier and later times from the progessive group that was their common progenitor, so that both in their successively more evolved modes of reproduction, and in their divergent directions of evolutionary change, the classes still inform us of the progress continuously advanced by that single group, which gave to each class in turn its genetic inheritance and consequent destiny.

Moreover, when they are taken in their sum, the classes inform us that the progress was always an advance to better modes of reproduction and regeneration, and never just a shift to new sorts of external adaptation; for it is by a comparison of their broad modes of reproduction, and not by way of their quite unclassifiable types of adaptive development, that vertebrate species are brought into their several classes.

The progressive group is accordingly not to be contained within any class. It was never just a jawless fish species, it was never a reptilian species, it was never a mammalian species. Indeed, in view of the fact that human populations are regenerated by group processes which bear but little resemblance to those which regenerate plant and animal species, it is questionable whether the group progenitor to vertebrate life was ever just an animal species. The history of that group is quite unique—it has never been told! Historical science is unlike all other science, in that its description of past occurrence may yet have to be radically revised in the light of future occurrence. What man is yet to be remains as yet unknown, and our histories of vertebrate progress, animal progress, living progress, and cosmic progress can never be finalized.

It follows that there was never any descent of man from mammalian species, nor of mammals from reptilian species, nor of reptiles from amphibians, nor of any class from another class. The narrow line of human descent was always the route of a progressive ascent to man, and it would really be closer to the truth to speak of a descent of vertebrate animals that was caused by their divergence from the line of vertebrate progress that ascended to man. Draw an ascending line to represent the direction of vertebrate progress, and let eight other lines diverge and descend from that one line to represent the descent of the vertebrate classes—a descent that was not only of species but of regenerative power.

However, the vertebrate classes were not retrogressive, nor even wholly unprogressive. Each of the class histories exhibits a minor progress that was just its own. The progenitor group was like a firework that launched eight rockets each soaring higher than the last. Reproductive character, too, is individually differentiated, and this necessitates a certain amount of evolutionary change of reproductive mode. Within each class there proceeded a continual replacement of less evolved and less viable species by stronger and more durable species. But, as we have seen, those minor progresses did not always reach far enough, under the pressure steadily intensified by animal increase, to save the classes from total or near extinction.

To what then, in summary, does the fossil evidence testify? Its direct evidence is of the several class histories. Here it tells us of the successively more evolved forms that have differentiated the classes since their entrances into the fossil record, and it also indicates something of the minor progresses advanced by the classes. Its indirect testimony is of the vertebrate progress itself. This was a sustained advance toward the reproductive process and regenerative power that would one day give to mankind an unassailable dominance over whatever lives on earth. That was truly a life-giving power, never to be defined or abused as a death-dealing power, intent upon the fabrication of weapons with which to destroy life. It gave life to the vertebrate classes; it must yet secure life to whatever lives on earth.

It is pleasant to be again in the company of Linnaeus, Lamarck, and Charles Darwin, those truly gifted thinkers whose perception of the living process advanced the progress toward an authentic life science, one that is proper to be the study of a modern age—of Linnaeus, first to recognize the processes of group regeneration that must forever remain the determinants of planetary life; of Lamarck, whose attribution to species of an inherent vital power never deserved the calumniation it received; of the younger Darwin who finally broke down the barriers that had prevented the advance to an evolutionary science; and of the mature Darwin who first advised us of the progress of reproductive character and behavior that was to make us human.

V

The Nature and Direction of Progress

During this century geologists have turned away from the conception of a catastrophic earth history to a belief that the physical environment of life has remained much the same since life began. Nevertheless, the long persistence of some species is still often said to have resulted from their occupancy of so-called "niches," these providing shelter from the pressures of environmental change.

The truth is that any species so harbored would pay a high price for its lodging. It would soon come to be so closely and uniformly adapted to its comfortable habitat that it would prove quite unadaptable to the new conditions that would sooner or later inevitably arise. Too uniform adaptation to any one sort or one set of conditions must be the usual cause of species extinction rather than a cause of species persistence.

Enormous extinction has marked every epoch, regardless of notable changes in the physical environment. There can be little doubt that it has resulted from the constant increase of life, which steadily strengthened the biotic pressures borne upon every species by other species. Generally speaking, species must have persisted just so long as they were able to regenerate themselves under these stronger pressures. No accession of regenerative power could come to species from the environment, the sole service of which is to destroy what is little adapted to its habitat. That power had to be generated from within the species themselves, and it could be gained only by means of some advance to better modes of group regeneration.

One way in which survival power might be augmented was through an advance to higher fertility. Because living populations are individually differentiated, there is always some selection of what is more fertile and fecund. Simply to increase the number of progeny is to widen the spectrum of diversified adaptive character presented to the environment, and thereby increase the likelihood that more individuals will come through growth to reproductive maturity.

There is consequently to be observed, more generally in the less evolved groups, a notable advance to modes of reproduction characterized by high fertility, some species exhibiting a truly remarkable fecundity. An individual oyster, for example, may produce half a billion eggs, with a statistical probability that just two of these will hatch into progeny able to survive to reproduce their kind again.

Quite evidently, this is not the sort of progress that gave rise to the higher animals which point the way to man. That history exhibits instead a decided trend to lower fertility, this being carried so far that the animals finally deliver progeny only one at a time and produce only a few in their lifetime. We are accordingly obliged to look elsewhere for the sort of progress that brought the more evolved animals to their high estate.

It is to be noted that in most or all living groups the rate of mortality is at its highest during the initial phases of growth, thereafter tapering off. Half of a new generation may perish during the first tenth of the growing period, a third of the survivors during the second tenth, a fourth of what is left in the third tenth, and so on, until in the last tenth of growth only one eleventh of the survivors perish. A little arithmetic shows that if 50% of the progeny produced in this group come halfway to maturity, more than half of these should complete the route. We are thus informed that any advance which prevents or moderates the holocaust of initial growth must be very effective indeed.

But what sort of an advance might this be? It would have to be one that by some means removed a growing generation from the first brunt of the selective pressures to which it would otherwise be fully exposed. If this progress of regenerative mode were then sustained, it would gradually remove from those destructive pressures an ever longer period of growth, and it is conceivable that such progress might at last withdraw a growing generation from all external pressures. If it did reach so far, that generation would need neither to be closely adapted to existing conditions nor adaptable to changing conditions in order to reach reproductive maturity and produce a next generation.

A progress of this sort, if it were carried to its full consummation, would draw a sleeve or muffler up and around a growing generation, until it eventually allowed each new generation to reach maturity and produce its successor irrespective of environmental conditions. It would completely insulate a living group from external conditions, so that this group would no longer be a species such as was described in the opening chapters of this book. It would be something quite unique and **sui generis,** not to be contained or confined within any classification of species.

Incredible as it may seem, to reflect upon this progess of regenerative power, and to obseve its large consequence as this is displayed in the long evolution of animal life, is to be convinced that it does in literal fact delineate the direction of evolutionary progress, which hindsight shows to have always been directed toward man. There has been long debate and much disagreement upon what most surely differentiates man from the animals, but not debatable and quite indubitable is the fact that man differs from all else that lives in his complete immunity from the selective pressures that forever bear upon other life. The advent of man necessarily established a wholly new realm of living nature.

Given this better understanding of the nature and direction of evolutionary progress, it is possible to reach an intelligible description of evolutionary history, for the evolution of life is primarily an unfolding of the large consequence of that progress as it is displayed first in the invertebrate phyla and later in the vertebrate classes.

Before there was cytoplasm or any emergence of cellular life, there may have been gene clusters left naked to the elements, but even the minute protozoa not yet advanced to cellular anatomy enclose their genic nuclei within an envelope of protective and nutritive plasma. Although still noncellular, protozoa are by no means simple organisms. Many have quite complex anatomies, and some have their own organs of locomotion. However, next to nothing is as yet known of protozoic history, and we must therefore go to plant and animal life for illustration of the early progress.

Because the earth was thought to be indefinitely old, while the legible fossil record goes back only some half-billion years, life was long supposed to be a rather recent phenomenon. Today, however, the age of the earth as a planet is estimated at only three or four billion years, while a few fossil traces are found to suggest that life existed perhaps half that time ago. A gifted paleontologist has written that the progress toward the amoeba must have covered a period longer than that from the amoeba to man. Life may be half as old as earth, or even more than that, as there appear to be minerals possessed of a power of self-reproduction that is usually taken to be definitive of living matter.

It was also believed, because plants are so obviously less evolved than animals, that the plants originated first and then gave rise to the animals. How or when multicellular life arose out of protozoic life, or how the plants and animals first diverged, no one as yet knows, nor has it ever been clearly stated just what it is that leaves the plants so obviously less evolved than the animals, although this is one of the broadest differences to be found in living nature.

Plants and animals alike testify to the long progress of regenerative mode that gave to initial growth a lengthening insulation from selective pressures. Cellular organization supported an advance to larger bodily size, and with this came an increasing specialization of cellular tissues for different functions, one of these functions being that of reproduction itself. The reproductive cells could now be enclosed within and nourished by the tissues of reproductive organs, and this opened the way to an evolution of new modes of regeneration, supported upon lengthening phases of parentally protected embryonic development. With this advance to larger body size and more complex anatomy there went a subsidiary evolution of adaptive character better fitting individuals to their particular habitats. (The rapid expansion of certain species into large groups of widely diversified species is usually attributed to some advance in anatomical complexity, the nature of which is left undefined, but it was necessarily an advance which bettered the reproductive mode.)

The plant seed encloses the new growth within a covering of nutritive meat which may in its turn be protected by a hard rind or shell, to insure that for a few days or weeks the seedling plant does not meet excessive moisture or drought, predatory bacteria of the soil, parasitical fungi, and other destructive agencies. The seeds may be equipped with wings or threads to make them airborne, or be made palatable and visibly or aromatically attractive to animals that will devour them and deposit the reproductive cell elsewhere, all of these devices serving to carry the seeds to new habitats where their growth will not be too congested.

The motile animals, of course, offer innumerable and much more remarkable illustrations of this regenerative mode that enables a mature generation to provide parental protection for its immature progeny. Spiders and insects may carry their young upon the maternal body, breed them in holes in the ground or under the bark of trees, hide them in nests or swathe them in spittle or cocoons, or rear them in capacious hives or subterranean labyrinths with nurses to foster them and soldiers to guard them. Fishes seek safe spawning grounds in pools lying above the rapids or falls of torrential streams, they deposit the eggs within submarine caves or under kelp beds or even under jellyfishes, or they may carry the young in their mouths for some time at risk of parental starvation (as does the male catfish, which may incubate several score of marble-sized eggs in his mouth, and later provide safe asylum there for the four-inch hatchlings). Most familiar is the parental diligence with which the birds may guard, foster, and educate their progeny. It has been observed that in some species many of the parent birds die shortly after the fledglings leave the nest, their thousands of flights made to forage for their progeny having left them too exhausted to fend for themselves.

Plants and animals accordingly both testify to this long evolution of parentally protective regenerative mode, but in the case of the plants the evolution proceeded very differently from animals, and to learn what left it so shortly truncated must cast some light upon the course of animal history. Although what initiated that ever-widening separation between the plant and animal kingdoms remains obscure, there is no lack of knowledge of its fateful consequence.

This was to make of the plants a permanently and incorrigibly sessile group, ineradicably rooted to the soil in which they find their major sustenance, whereas the animals, although they likewise long remained predominantly a sessile group, apparently had always open to them the way to motile habit. This potential capacity is indicated by the squids and octopi, which carry within themselves a vestige of shell attesting to their descent from ancestral mollusks, but which evolved a remarkable—among the animals quite unique—means of locomotion by jet propulsion.

As indicated above, sessility is properly speaking an adaptive character, one that serves to keep the plant tied to its mineral sustenance, or that keeps the sessile mollusk in the path of water currents that bring it nourishment. Mention was made of some of the ways in which plants secure a wide dispersal of seed. Since the advent of the insects upon land the most important feature of plant evolution has been the symbiosis that ties the plants to their insect pollinators. But all of these ways only increase the dependence of plant species for their vital gene exchange upon agencies external to themselves and out of their control. For the insects and other animal pollinators, that symbiotic relationship (if it can properly be called such) is only a form of behavior that provides sustenance—it is only an adaptive character that can be at need replaced. But for the plants it is an essential element of their mode of group-regeneration. It might almost be said to leave the plants something less than true species, as may be indicated by the frequency of hybridization among the plants and by their recourse to degenerate non-sexual modes of reproduction.

The plants bring to mind the biblical Esau who sold his birthright for a mess of pottage, for the exchange of a power of self-generation for some adaptive advantage that can secure only a brief span of individual existence must always be a sale of the birthright of life.

It is accordingly only in animal history that evidence can be found of the later progress to man. Least evolved of the animal groups are those containing sponges and corals. These sessile and colonial animals secrete hard skeletons which hold the small individuals in close proximity, presumably to make possible the exchange of genotype. They are followed by four groups of wormlike animals, and by the lampshells and a group containing starfishes and sea urchins. These earlier invertebrates tell of a progress leading to a threefold differentiation of tissue layers encosing a body cavity. The lampshells also initiate the advance to bilateral symmetry that would be carried much farther by the more evolved groups. Then came the mollusks, or true shellfishes, which today comprise some 80,000 species. They left an unusually complete fossil record which makes their history the best documented of all living groups. One of their orders in particular was long the subject of intensive study, so that it still provides the clearest ilustration of the process of expansion and decline earlier described.

Some 200 million years ago, at the time of the Permian-Triassic crisis, when great mountain-building made uninhabitable many regions formerly densely populated, there came a congestion of animal life that aggravated all of the biotic pressures exerted by species upon species. The ammonite shellfishes were then a widely expanded group comprising twelve families of related species. Of these families all but one were wiped out! Only a few species remained, and these belonged to just one ammonite family and even possibly to just one ammonite genus. However, these few surviving ammonites now underwent such rapid and wide multiplication that within thirty million years they were again expanded into a great order, containing no fewer than twenty-three new ammonite families in place of the eleven extinct ones.

Exhaustive study of the excellent fossil record left by this group has never revealed any notable difference of externally adaptive character, such as might be taken to explain this virtually complete replacement of the ammonite order by the few species that arrived so much later at their phase of wide expansion. Much the same selective pressures had worked upon both the earlier and the later ammonites. Nor could it have been any particular local change establishing some special sort of readaptation that brought the earlier ammonites through wide expansion to near extinction, or that let a few species survive to be carried to even wider expansion. Ammonite history does not have a happy ending, telling us how the restored ammonites lived blissfully forever after. The later expanded ammonites had no sooner fully restored their great order than they too came to decline, which this time was not to stop short of total extinction.

Is this an unusual history? Or is it only an unusually well documented instance of the process of replacement that is a universal feature of the evolution of life? Whence came that singular species that arrived so belatedly at its expansive phase, when it established a new genus of species destined to be so much multiplied that they established twenty-three new families of cognate species?

That remarkable species was the terminus of a line of descent which necessarily went all of the way back to the single species progenitor of the whole ammonite order, as did, of course, all other lines of ammonite descent; but it evidently was and always had been the strongest, the most firmly integrated, and the most widely adaptable species of its order, probably destined to outlive all the others. Strong species are not prone to population increase. They are not like the weaker species that break off from them and then multiply into new genera, but they persist obscurely and safely among those weaker descendant species. Must we not conclude that the singular species which restored its ammonite order was none other than the species originally progenitor to the order?

The regenerative powers bequeathed to species are not unlimited. No species is invulnerable; and all species, too, are mortal, each of them carrying within itself the seeds of its own demise. Even the strongest species perisists only by virtue of its exceptionally wide adaptability to environmental change, and it owes this adaptability to a capital fund of diversified genotype which is steadily depleted as the species becomes uniformly readapted now to this and now to that local condition. The increasing pressures to which the later ammonites were subjected finally brought even that strongest progenitor species to the close and uniform adaptation that precludes wide adaptability, and finally this species too, perhaps as the last of its order, was splintered into extinction.

Today, the accelerated expansion of agriculture and industry subjects species to selective pressures more destructive than any imposed upon plant and animal life in all the long past. If species are still to persist, it will be because man, now their destroying angel, becomes their redeemer;

and should not man, whose remote or nearer ancestors have always been the creators of planetary life, now knowingly assume responsibility for its continuance?

Last and most evolved of the invertebrate groups is **Arthropoda.** This great phylum is named for the jointed feet or legs that most obviously differentiate its species from all other invertebrate species. Interestingly enough, this name does not distinguish it too clearly from the vertebrates that were to follow, for the term 'vertebrate' (derived from Latin **vertere**—to turn) also refers to the jointed limbs of the animals comprised within that group.

The group **Arthropoda** has become by far the largest, most widely distributed, and most remarkably diversified of all animal groups. It is believed to comprise no fewer than 800,000 species, ten times the number of mollusk species, and twenty times that of vertebrate species. It comprises, indeed, about four-fifths of all animal species now existing on earth. It includes crustaceans and spiders along with insects. It soars in the air, it covers the earth, it plunges deep into the waters under the earth. Its small crustaceans provide a substantial part of the plankton that is the ultimate sustenance of most marine fishes. In its most evolved orders it includes the social insects, which both in their modes of group regeneration and in their social economy offer a parallel to—or is it a parody of?—human society.

And the group still expands! An occasional scientist surmises that the future may well belong to the insects! This conjecture may promise interesting material for a writer of scientific fiction, but it is not an hypothesis that can be seriously entertained by anyone aware of the true nature of evolutionary progress. The insects, despite their advance to something like social habit, have no resources that would let them compete with mankind, whose mode of regeneration forever immunizes him from the environmental pressures that bear upon all other living groups. Of course, if it is permissible to revert again to the moral of this story of evolutionary change, it is not impossible that man should yet destroy himself, either through his own lethal weapons or by his failure to assume responsibility for other life, for the plants and animals which he seems at present so willing to destroy must remain a necessary condition of his own existence. Man might conceivably live on bread alone, but he won't last long on pure chemicals.

VI

Vertebrate Progress

For evidence of animal progress one must go first to the invertebrate phyla, these being thought to be progressively more evolved, and after them to the eight vertebrate classes, these indicating as many consecutive stages of the vertebrate progress. For our identification of all of these animal groups we may confidently rely upon the classification of species, but neither this nor the fossil evidence tells us when or in what order the groups originated. The classification of species does not disclose the order in which its largest categories were established, and the fossil record leaves all origins obscure. We do learn that the vertebrates came last of all the phyla to wide expansion, but the invertebrate phyla are all represented among the oldest fossils, so that they might conceivably have originated in short succession and come down the ages side by side.

Nor is a diligent study of the many groups as they observably exist today at all conclusive concerning their origins. For example, there is wide and increasing disagreement as to just which invertebrate group gave rise to the vertebrates. Most scientists believe it was the crustaceans included in the great group **Arthropoda,** but some hold it to have been a shellfish species, and others a species of annelid worms, that impossibly found itself producing vertebrate progeny. This uncertainty casts into doubt any hypothesis that would list the many animal groups in the order of their succession, with each group in turn giving rise to a somewhat more evolved successor.

In view of this uncertainty about invertebrate history, it is surprising to find well-nigh unanimous agreement about the order in which the progressively more evolved vertebrate classes supposedly arose, each to complete its due quota of the vertebrate progress to man, for here, likewise, neither the classification of species nor the fossil record provides any evidence of the class origins. A comparison of the classes does convince us that probably the largest part of vertebrate history was aquatic history, the plants, shellfishes, and insects having come earlier to land. But why has it been so widely and unquestioningly believed that the vertebrate ascent upon land was an amphibian achievement, when all of the evidence indicates that the amphibians are less evolved than the bony fishes which are taken to have given rise to them?

The oldest vertebrate fossils show the bony fishes and the amphibians existing side by side, with the bony fishes arriving later than the amphibians at their phase of wide expansion. Moreover, both groups to begin with were freshwater groups, the bony fishes coming only later to sea, and it is believed that the late expansion of the bony fishes was one cause of amphibian decline.

The amphibians certainly do seem at first sight to resemble the reptiles much more than they do any of the fishes. In place of fins and gills they have legs and lungs, and at least some of them do live on land. It was perhaps inevitable that they should have initially been regarded as a first terrestrial class, one which first came to land and later gave rise to the more evolved reptiles. However, the earlier bony fishes did have lungs, these being later lost, and there is much more compelling evidence to preclude any serious entertainment of that so generally accepted but basically unsound description of vertebrate history.

It is true that some amphibians are able to live on land and even produce progeny on land; but, although porpoises and whales spend all of their lives at sea and likewise produce their progeny at sea, they are not included among the aquatic vertebrates, the reason being that their mode of reproduction requires them to be included among the terrestrial mammals. Where those sea-going mammals retain their terrestrial mode notwithstanding their development of aquatically adaptive anatomy, the amphibians keep their aquatic mode whatever may be their external adaptation to terrestrial conditions. The classification of species must always remain our largest and surest compendium of biological fact and our most reliable evidence of the genealogy of life, and it is such because it classifies species always and only by their similarities and differences of reproductive mode, and never by their often convergent patterns of adaptive character—just as Linnaeus decreed when he first began the classification.

A competent zoologist has written: "The amphibian is still chained to the water. . . . In the water it is born, to the water it must periodically return. . . . It is, in reality, little more than a peculiar type of fish which is capable of walking on land."[3]

Amphibians may lay their eggs in the wet crotches of trees, in the wet earth under stones, and in other sufficiently moist places, although most of them still breed in the waters of lakes or streams, their progeny hatching out there and completing an initial phase of growth under water. In one amphibian species the male wreathes the strings of fertilized eggs about his own moist body and goes into hiding, emerging at night to douse the eggs in ponds. In another species the male pushes the fertilized eggs under the skin of the female, where they incubate. These are admirably domestic modes of regeneration, but they still leave those amphibians indubitably aquatic, and forbid their inclusion among the truly terrestrial vertebrates.

Many of the bony fishes travel in large schools, and these must owe their cohesiveness to some means of intercommunication, similar to that observed among birds that feed and travel in flocks. Behavioral adaptation of this sort points to the progress that would eventually give to human societies their communicant character and their cohesive structure, and it goes beyond anything found in amphibian life. Any ichthyologist can offer almost inexhaustible evidence going far beyond what is needed to remove any doubt that the bony fishes are much the more evolved class.

A final testimony may be drawn from the fossil evidence, which shows most of the early amphibians to have been insectivorous. The much earlier expanded insects were necessarily the major sustenance of any carnivorous vertebrates that first came to seek food on land, and aquatic animals that had snapped up insects flying above the water might well have come to pursue their small prey upon land. It was by a purely adaptive innovation that the amphibians were given legs and perhaps lungs in place of their former aquatic anatomy.

The newts, frogs, and toads familiar to us today little resemble any of the bony fishes that are said to have given rise to them. They are the descendants of a few species that came through the Permian-Triassic Period of great mountain-building. These species had successfully met those rigorous pressures, and also those later imposed by the expanding freshwater fishes and the later expanding reptiles. They must have always been the oldest and strongest species of their decimated class.

Strong species, just because they are well integrated and widely adaptable species, are not prone to population increase and thereby likely to be carried to wide expansion. The several species ancestral to the now existing amphibians, or possibly the single species that was multiplied into them, must have persisted quite obscurely through the Amphibian Age (which is to say 100 million years and more).

That single species may well have been the species progenitor of the whole amphibian class, still obscurely persisting while other species were much multiplied, and only later, after those descendant species became extinct, would it finally come to its present modest expansion into the amphibians now existing. What the paleontologist learns from the fossil record shows these latter to differ markedly from the early expanded and now extinct amphibians. This history should convince us that the amphibians, originating and widely expanding before the bony fishes, neither descended from these nor gave rise to them.

However, if the amphibians never really came to land at all, but were and remain a nonterrestrial class, it has now to be ascertained how the terrestrial vertebrates did make good their advent upon land.

Vertebrate history goes back to a far time, 100 million years and more before the great mountains were built, when flatter continents were washed by shallower seas, these seas being less salty than they are today. Sluggish waters seeped through vast swamps bordered by wide tidelands fretted with many a brackish lagoon. It was in those tepid tidal waters, laden with mineral matter and later also organic matter washed down from the land, that proceeded the first large expansions and deep replacements of animal life. The wide and shifting littoral especially swarmed with aquatic life, predaceous species exerting upon one another—and always first and foremost upon initial growth—selective pressures steadily hardened by living increase and sharpened by newly evolved types of predatory adaptation. Animals could grow fins and legs, gills and lungs, or both or all, their external adaptation could be at once aquatic and terrestrial, so long as they retained their fully aquatic mode. It was only later, when the mountains came, that coastlines became permanent and clearly defined, so that external adaptation had usually to be either aquatic or terrestrial but not both.

It was under the earlier conditions that one or more vertebrate species came to find upon dry land the breeding places that might remove from increasing aquatic pressures a brief initial phase of growth. There had to be longer and better embryonic development, productive of large and tough-skinned eggs that could be incubated out of the water. But this and other conditions were progressively met. To begin with, the period of land-tenure would be brief indeed, the fishes coming to shore only to deposit their fertilized eggs, as do the great sea turtles still, and only very slowly, as sexual and natural selection worked together, would the period of terrestrially insulated growth be gradually lengthened. But the time came when the incubated progeny would no longer at once race down to the water that was still their natural medium (as do new-hatched crocodiles to African lakes today, if they can get past the insatiable marabou "with his hangman's eye" that waits for them at the water's edge). The new terrestrial mode of animal regeneration was steadily bettered, the return to water was longer deferred, an ever larger fraction of the period of growth was spent ashore, until at last the whole growth was completed out of the water, and the epochal ascent upon land had been achieved.

But some of these newly terrestrial vertebrates would never relinquish their adaptation of aquatic type, these species breeding on shore but still gaining their major livelihood under water, and there were even those that after long sojourn upon land would wholly return to the water, making of themselves fishes again in all that was not their now indelible reproductive mode. They would live and breed at sea, but do so as reptiles and mammals and not as scaly fishes. (Do we not ourselves, whenever we meet surf and sand, at once succumb to that old sweet deep nostalgia, and let wave and weed and tidemark whisper to us of whence we came, and of where we were so long held fast, one foot on shore and one in sea? "Come unto these yellow sands!" Darkly but deep in human memory stirs that ancient passion, to which many of us are faithful still after our fashion.)

Is this history, or just fantasy? There exists today a species of bony fishes that has been brought down the many periods and epochs of animal history as if just to attest its fidelity to fact. These are the grunion, whose present habitat lies off the Pacific coast of California in North America.

These small fishes have astonishingly synchronized their breeding processes with seasonal high tides in order to provide safe lodging for the fertilized eggs. On certain moonlit nights the millions of conjugal pairs swim through the high surf to shore, and let themselves be swept as far up the beach as the waves will carry them, when they wriggle still higher up the wet and sloping sand. The female, almost erect, digs a hole in the sand with her tail and lays her eggs in it. The closely attendant male at once fertilizes the eggs and buries the nest, and the mated pair then squirm their way back down the beach to ocean, all of this having happened in less time than it takes to tell, or at least to write. One wonders whether moonlight did not weave a fantasy of glinting sea and shining beach and industrious fish, but the nest is there, to be left high and dry in the sun-warmed sand by the receding tide, and in just two weeks, when there comes the next high seasonal tide, the eggs are incubated and the progeny leap into the water of the first wave that wets the nest. These tiny grunion are now fair game for whatever forages in the surf, but the two most hazardous weeks of their precarious growth are safely behind them, thanks to parents who had learned their astronomical lore from God knows what indulgent star, and we may be sure that some of them will come again to that wide and shining beach that was their womb, as grunion must have come these many many million years.

But not all of those gallant grunion make their slithery way back to sea. Today, when they hear of a grunion run, folks come from nearby towns and cities to scoop the small fishes into pails, for they make good eating, and, for a week or more after a run, the tidemark kelp is littered with the dried-out remains of the many grunion that gave their all for love. These losses the grunion may survive, and may no oil slick or other industrial pollution of our coastal waters put an end to this miracle of parentally regenerate life.

Once the great transition to terrestrial mode was completed, vast continents rich in insect and vegetable sustenance and as yet empty of vertebrate predators invited habitation, and, under such low selective pressures, the tempo of evolution was temporarily uncontrolled, for there could be large increase without great uniformity of mode and type. The result was the wide and remarkably diversified radiation of terrestrial vertebrates that began about 250 million years ago and ushered in a Reptilian Age, comparable in its exuberance with the great explosion of insect life come earlier to land, and productive of a panorama of reptilian "forms" which in its pattern of adaptive types would be rather closely paralleled by those of the mammalian life that was to follow. Some fifty million years after the reptiles were fully expanded there came the crisis of mountain-building, and the reptiles along with all other animal groups were much reduced, but they made admirable recovery, more than restoring their great class, and maintaining their undisputed dominance through yet another 100 million years, when they suddenly fell upon world-wide and permanent decline.

Nowhere, surely, is there to be found so spectacular an illustration of expansion and decline as was that of the reptiles. They established sixteen orders, and these contained small, medium-sized, large, and enormous reptiles. There were reptiles that walked or ran, winged reptiles that flew, finned reptiles that swam the seas, and limbless reptiles that snaked their way about. Yet all of these so diverse anatomies were modifications of one and the same organic structure, as was required by the mode of reproduction that came to all reptiles from their common progenitor. Even so they underwent some minor modal change, most of them remaining oviparous, but a few advancing to an ovoviviparous mode that retains the eggs in the maternal body and delivers them motile. But they still remained bona fide reptiles, they could not really become fishes or amphibians—or mammals! Some reptiles are commonly said to be amphibious, and others are called purely aquatic, but this description confuses their externally adaptive character with the broad mode of reproduction which alone brings them into their reptilian class. The leopard may change his spots or shed his skin, but he cannot rid himself of his reproductive mode; he must still remain a mammal.

There have been many conjectures as to what brought the reptiles to their catastrophic decline. It is not unusual to find such decline depicted as a result of pressures imposed upon a fully expanded class by a newly expanding class, but, in this case, that expanding group would have had to have been the mammals, and it is difficult to imagine tiny mouselike mammals—for such they still were in those days—assailing gigantic dinosaurs and formidable tyrannosaurs in gladiatorial combat and coming off the victors. (It is seldom remembered that selective pressures must always primarily work upon growing animals, what happens to adult animals that have produced a new generation being of small evolutionary consequence.) It was accordingly convened that the mammals did not displace the reptiles and only expanded widely after reptilian decline, which left the reptilian habitats untenanted and open for occupancy. Dark allusions have been made to protracted cold, to unusual heat, to prolonged drought, or to some other climatic or physical change.

There is no evidence of any such change. The great crisis was long past, the mountains were built, and other physical conditions were much what they are today. A zoologist well acquainted with snakes has propounded the theory that the great reptiles were prevented by long-continued heat from their production of spermatozoa, so that they could no longer breed, but the snakes and lizards and crocodiles still bred, and there is no record of that unseasonable heat. It had to be the mammals, large or small, that actually destroyed and displaced the reptiles—either alone, or with an assist from the birds.

As a class, the reptiles bring little care to their progeny, and, what is more important, they are cold-blooded. Warmbloodedness is too often said to be only the overt sign of a high rate of metabolism, but faster metabolism then conditions a better production of cellular tissue that may toughen and improve every organ. The cold-blooded animal digests slowly and must and can subsist upon a relatively low intake of food, but this is of small advantage in a predatory world, as it leaves the animal with little available energy for attack or defense. Moreover, the cold-blooded animal is somewhat dependent for its motile response upon a fluctuating provision of daily and seasonal warmth, so that its behavioral response is often delayed or even prevented. This placed an insuperable handicap upon the small and lethargic reptilian young, leaving them often defenseless against warm-blooded predators, also small but extremely active. Some shrews daily ingest several times their own weight of animal food.

However, the supreme virtue of warm-bloodedness lies in its promotion of efficient reproduction. In the birds and mammals it is the warm-bloodedness of the parent that quickens embryonic development, and that after the progeny are hatched or born brings them through their initial growth. Only the rapid metabolism of the birds makes possible their quick production of eggs so enormous relatively to the small bird that produces them. Only warm-blooded brooding upon the eggs ensures their incubation, and it is often parental warmth together with the well-made avian nest that keeps alive the nestlings whose small and still naked bodies quickly lose their own body heat. We are wont to see in the feathered covering of the birds only a condition of their power of flight, but its first and largest service is its betterment of parental nurture, much as is that of the fur or hair of the adult mammal.

It is to be observed that large birds removed from or secured against animal predators tend to lose their power of flight, the progeny here being safely incubated upon the ground. We must conclude that bird-flight likewise originated as a reproductive character, its initial service having been to remove the nest, earlier built upon the ground, from animal predators. Warm-bloodedness and bird-flight are equally features of reproductive mode and group regeneration, and the former is possibly the most consequential advance made in animal history.

As the birds and mammals expanded, their carnivorous species turned to reptilian food. It was not by any assault upon the adult reptiles, gigantic or small, that they emptied the reptilian habitats. These warmblooded species found their sustenance in the unguarded reptilian eggs and in the defenseless reptilian young. The three surviving reptilian orders, these being the turtles, the snakes and lizards, and the crocodiles, descend from species certainly the best regenerated of their class, which did afford some protection to their progeny. Consider the American alligator that is so cruelly harassed today!

"The female builds a high conelike nest of plants gathered from the neighboring marsh. She deposits her eggs and proceeds to guard the nest until hatching time sixty days hence. During the incubation period, the female clears the plants from an area round the nest, creating a nursery pool for the youngsters when hatching time comes. At one end of the pool a hole is dug into the bank leading to a dry cave which will be home to the brood for many months, as long as two years in some instances. A few hours or a day before hatching, the baby alligators begin to grunt within their shells, and when the mother hears the sounds she climbs the cone and claws away at the debris until the eggs are exposed to view. She is careful not to touch them, however. Meanwhile she responds to the gruntings with deeper-toned growls of her own. When the hatchlings begin to appear at the rim of the nest the tempo of her grunts becomes more rapid, and she leads them down the slope to the pool ready to receive them. During her sojourn with the babies she feeds them by crushing a duck that has fallen wounded from a hunter's gun, or a young heron that has fallen from its nest. The carcass, thus macerated, is presented to the brood to be devoured hungrily. Alligator hunters frequently designate an alliator brood as a one- or two-year brood, thus emphasizing the very strong responses of a mother that remains to guard and feed her brood for periods longer by many months than most avian or mammalian mothers are known to care for their offspring."[4]

That the birds originated after the reptiles but before the mammals is indicated by the scaly covering of their legs and feet (the feathers of the bird are a cellular modification of those scales). This conclusion is not too well supported by the fossil evidence, but the latter does indicate that the two classes both had long obscure histories before they came to expansion wide enough to bring them into the fossil record. The light and fragile bones of the birds make poor fossils, so that we should not expect their early history to be recorded as fully as was that of the heavier boned mammals. In much of their behavior—in their exuberant sociability, their fearlessness and tolerance of man-made environments, their melodic and often communal song and occasional mimicry of human speech—the birds may seem closer to man than do many of the mammals. However, the extreme furtiveness of most mammals today may be partly a consequence of the pressures exerted by human expansion. It is apparent that their oviparous mode has always set very definite limits to the progress of the birds, as did also their evolution of winged flight, this keeping them small and light (it is only the small passerine birds that really flourish), and thereby precluding any advance to embryonic development comparable in length and efficiency with that of the larger mammals. Flight could be relinquished in the interests of larger size and correspondingly longer growth, but the largest bird must still be incubated, and embryonic growth within the egg cannot compare with that within the womb. Nevertheless, when we come to review the behavioral evolution of the animals, we shall be able to record a very substantial achievement of the well-nurtured birds in the advance from the purely instinctive behavior of the lower animals to the intelligent behavior of the higher animals and man.

For our evidence of the last 150 million years of evolutionary progress we must go to mammalian history, which alone tells us something of the advance to the regenerative mode bequeathed to earliest man. Where the animal phyla point to the earlier progress, it is in turn the vertebrate classes, the families of the most evolved mammalian order which contains the primates, and the primate families the most evolved of which contains the hominoids, that may direct us to the successive stages of the later progress.

The earliest mammals of which we can claim to have authentic knowledge are monotremes like the platypus and echidna, and these animals still lay eggs, although they also have milk-secreting glands which point to the modal progress to come. Considerably later must have originated the marsupials, suitably named for the maternal pouch to which the tiny newborn progeny at once crawl for further protection and nourishment. (The baby possum is no bigger than a bee.) And very much later must have appeared the true mammals, so emphatically differentiated by their efficient placental gestation of new growth. Embryologists are agreed that only in this fully mammalian mode of reproduction is provided that superlative metabolic association of mother and fetus which brings to embryonic growth its optimum conditions. And it was this reproductive interadaptation between consecutive generations, we all know, that was progressively bettered during the last 100 million years of mammalian history, steadily to lengthen embryonic development to the point where this had to be followed by much longer postnatal nurture. Still later, the time came, one or a few million years ago, when this reproductive mode had evolved into the incomparable domestic development that is a human family, that larger womb within which is generated the social process that has directed human history and that still must preserve mankind.

Vertebrate history is set off from invertebrate history by another great difference, and it is one that appears at first sight to be as disturbing as it is inexplicable. Among the invertebrates, we find that the most evolved groups are also the most widely expanded and most successful groups. These are **Mollusca** and **Arthropoda,** which together account for almost nine-tenths of all animal species existing today. Among the vertebrates just the opposite may be found—the more evolved a group, the smaller is its prospect of persistence! The mammals now decline while the birds still expand, and, among the mammals, the less evolved rodents do much better than the more evolved primates. Finally, among the primates, we find that the less evolved New and Old World monkeys do much better than the more evolved group containing the hominoids. This last group, moderately expanded twenty to thirty million years ago, fell at once into decline, and persists today only in four small groups of great apes. Finally, if there ever was established a genus **Homo** containing **Homo sapiens,** that group persists only in man, whatever else it may once have included having completely disappeared.

We seem to be shown a great parabola that graphically depicts the rise and fall of a capacity for wide and sustained expansion. The parabolic curve has its apex in the enormous and still expanding insect group. From that climax it now descends, slowly in the vertebrate classes, steeply in the mammalian families, precipitately in the hominoid genera, where it becomes plumb vertical and returns to base, no genus **Homo** being established at all.

Yet, why should this have been, how could it ever have occurred, when earliest man, most evolved of all that lived, could proceed through the fastest and widest expansion to be found in living history? We may well maintain that man is to be included within no animal category, large or small, and that he is of no class, ordinal, familial, or generic descent. The small group ancestral to man had never itself been just a species. It was of all groups always the most adaptable, this being assured by its preeminent mode of regeneration, and, while it inevitably underwent continuous readaptation, it never bought readaptation at some cost of its regenerative power. As it came down the earlier ages, its widening adaptability made it the progenitor of species capable of longer persistence and wider expansion. However, when the time came that it ceased to be the progenitor of animal phyla and became that of vertebrate classes, it had begun its advance to reproductive modes that gradually withdrew it from selective pressures, and it did not now need to be so widely adaptable in order to persist. This lessening adaptability, with its consequence in a steadily decreasing capacity for expansion and longevity, it bequeathed to the species that were multiplied into vertebrate classes, mammalian orders, primate families, and hominoid genera. It was only after it had completely insulated itself from all selective pressures whatsoever that it could be itself rapidly multiplied and widely expanded, to cover the earth in an explosive radiation that has never had its like since life began.

Vertebrate history, of all histories the most dramatic, does not lack its tragic episode. The later and more evolved mammals were given a dubious inheritance, one that fell short of the genetic capital assuring them of long persistence. Raised above their fellow animals in their closer kinship with the group from which they all had diverged, they remained still subject to the pressures those animals imposed. What is it—envy, reproach, or just incomprehension—that speaks to us from their eyes as they look back at us from their prison cells in our zoos?

VII

The Evolution of Behavior

Ethology, the study of animal behavior, has come very belatedly to its place among the descriptive sciences. "Study of the evolutionary aspects of behavior," an ethologist complains, "has lagged far behind the morphological study of evolution",[5] and another ethologist writes that, "if we had a record of behavior as complete as the fossil record of structure, this would leave as convincing a body of evidence as does the latter."[6]

The truth is that the paleontologist finds evidence in the fossil deposits of evolutionary change only because he comes to the fossil record with considerable anatomical knowledge gained from a comparative study of animal groups as these observably exist today. As soon as the ethologist gains a similar knowledge of animal behaviors as these may be observed today, he will find the fossil record just as informative to himself. Because the classification of existing species provides a perspective upon the genealogy of species, it provides a living record of the evolution of life which the fossil record may then supplement and confirm.

This living record of evolutionary history goes all of the way back to the primeval protozoa. It used to be thought that these microscopic animals reproduce themselves by simple fission, a single protozoon being multiplied by repeated division into clones containing innumerable protozoa. However, it was observed that protozoa of different clones actively conjugate, two protozoa merging, exchanging genes, and separating again, then to be multiplied into genically different clones. It was slowly realized that this "rejuvenating" process is in fact the protozoic mode of reproduction itself, and one that does not so widely differ as was thought from that of the multicellular animals or metazoa. The metazoon is really a tightly knit and complex clone, consisting of myriad cells all of identical genotype because they are produced by the repeated division of a single reproductive cell. But it is only for the incorrigibly romantic protozoa that love and death are one and the same awesome event. Protozoa, like metazoa, are sexually reproduced.

The digger wasp, appropriately named for its reproductive behavior, digs a hole in the ground, finds and paralyzes a caterpillar of a certain species, drags this fodder into the hole and lays an egg upon it. It then seeks foliage of certain sorts, brings this to cover the caterpillar, and finally fills and seals the hole. It may even construct several of these artificial wombs at once, running industriously from hole to hole. We surely cannot call this behavior unintelligent. The wasp's behavior shows remarkable intelligence, first of the conditions conducive to embryonic growth, and then of other species. Nevertheless, this is purely instinctive and unlearned behavior, for the larva which develops from the buried egg, if and when it reaches maturity, will approximately repeat the parental behavior without any instruction or example from its kind.

Instincts are genically transmitted patterns of behavior. Just as the genes bequeathed to an animal determine its cellular, glandular, digestive, and other development, so they may determine its behavioral patterns. All-important is the reproductive instinct that brings the mature animal to conjugation with those of its own species. Other instincts are genically inherited patterns of adaptive behavior.

Instinctive behavior is not to be confused with reflex action. The latter is not properly behavior at all. It is only the reaction of a body organ to some reagent, as when one's skin perspires in reaction to external warmth. Reflexes are of a fixed and stereotyped character, and it was mistakenly believed that instinctive behavior has this character too. But, it has been found, only the component elements of instinctive behavior, not the behavioral pattern itself, are so stereotyped. Instinctive behavior is not inflexible, it is not fixed. It may be interrupted and resumed, it can be improved with use, it may be continuously readjusted to a moving or charging stimulus, as occurs in every animal pursuit of moving prey.[7] We should not disparage instinct, for it is still there, perhaps, rather than in acquired knowledge, that wisdom may house its ancient oracles.

The instinctive behavior of the digger wasp described above is only one of the innumerable forms of reproductive behavior, often most ingenious, that make of the insects the most widely diversified of animal groups. Among the vertebrates some of the fishes display an impresive evolution of conjugal and parental behavior, more advanced than that of the reptiles and resembling that of the birds and mammals. It is not true that the motile fishes simply cast their spawn upon the waters to be fertilized as wave or wind may decree. The motile fish may closely follow and assiduously court the gravid female, lead and coax and maneuver her into the nest he has prepared for her reception, and repel by threat or bodily attack any real or imagined rivals. He may guard her from predators while she lays and he fertilizes the eggs, and long remain to watch over the progeny, to which by a fanning motion of his fins he directs a stream of oxygen-bearing water. Most dramatic and strangely moving can be the behavior of the breeding fish, whose cold blood seems in no way to cool a most ardent passion. The bony fishes appear to be no better externally adapted than are many of the less evolved fishes, and they quite evidently owe their dominant status to their superior modes of domestic behavior.

The stickleback fish whose mating behavior was just described makes use of the same fanning motion as a phase of his courtship of the female. This rather unusual behavior is a promise of readiness to assume paternal responsibilities, and, much as the cat's lifted paw or the dog's bared teeth conveys a threat, it is a significant instance of instinctive symbolic communication, already pointing to the progress that would later bring speech to man. Communication is effected through the use of some phase or element of behavior presented outside of its usual context, so that the part may symbolize the whole.

This instinctive behavior would appear to entail a causal inference on the part of the animal to which it is addressed. Causal inference is the ground of reason, possibly its very substance. Reason may be as old as instinct, and it must have arisen along with animal motility. The motile animal must in some degree be sensible of what lies about it. The sensation given through a sense organ, however, is internal to the animal, and it becomes perceptive only when it induces a causal inference to what lies outside of itself. Evidently the beginning of reason lies in sense-perception, for every perception, animal or human, entails an implicit cognition of causal relationship.

Even the tiny fruit flies, so useful to the geneticist because they breed so rapidly and are so tolerant of laboratory conditions (**Drosophila,** the lovers of dew, is the charming name the entomologist found for these frequenters of our garbage pails), symbolically communicate to others of their kind their group-identity, their sex, and their sexual condition by means of appropriate postures and movements. Among both insects and vertebrates, these being of all animals the most evolved, conjugal behavior is often implemented by some means of sonic communication. Crickets chirp, cicadas cheep, frogs croak, birds sing, gulls mew, cats meow, dogs howl, cattle low, and elephants trumpet, all to bring individuals of opposite sex, but of one and the same species, to awareness of each other. But it is only among the more evolved and highly domestic vertebrates, where long parental nurture follows upon conjugal association, that sonic communication came to implement the parental-filial association between consecutive generations.

Note was taken earlier of the grunting of young alligators, warning the mother of their imminent emergence from the shell, and we may now contrast the purely instinctive behavior of the digger wasp with the behavior of a vertebrate animal that nurtures its progeny. The first cheep of a chick, like its mechanical pecking of the ground that repeats the motion which broke the shell, is undoubtedly instinctive, as is also probably the answering cluck of the parent hen. However, once this ground of parental-filial communication is established, it becomes the support of a domestic association that thereafter directs and shapes behavioral development, this then importantly conditioning all further development of nervous and glandular tissue. The progeny instinctively seek and respond to the parent, and the mother bird, by modifying her call, gathers her brood together, leads it to and from feeding grounds, sends it hurtling to cover at the approach of danger, and steadily induces in the progeny the adaptive responses that better secure their survival in a predatory and dangerous world.

These features of nurtured growth, so familiar to us all, have never received the attention that is their due. It is quite evident that the unnurtured chick could not live, so that within a nurtured group survival ultimately depends upon types of adaptive behavior that are not genically transmitted, but nurturally transmitted, from each adult generation to the next. Here is an inheritance and bequest of adaptive character which, in sharp contrast to that genically transmitted, is directly made during growth. Within the nurtured group, survival comes to depend less upon inherited genotype than upon behavioral adaptation directly induced by parental nurture.

With lengthening nurture, learned behavior steadily displaced instinctive behavior as a condition of survival. The human infant must be weaned and taught to eat and drink, to stand and walk, to sleep and defecate, to speak and feel and "behave itself" humanly. Long ago Darwin explained changes of dentition and facial structure in primitive man as an effect of the increasing use of tools. The stone-age technique was nurturally transmitted, the dental character still genically transmitted down the generations, but the selection of what was behaviorally more advanced overrode that of what was anatomically more adaptive.[8]

We can now see that the advantage of learned behavior over instinctive behavior lies in its greater plasticity. An improvement of instinct due to use will not as a rule affect genotype, nor result in a change of heritable behavior, but, within the nurtured group, any new condition affecting individual growth will be reflected in the parental conditioning of progeny. A parent threatened by some new sort of predator or nourished by some new sort of sustenance will duly condition its progeny to flee the one and seek the other. One might say that where there is no postnatal nurture, all change of group character must be effected through the narrow bottleneck of the reproductive cell, whereas in the nurtured group it is the whole growing organism that can bring about such change.

All of the foregoing instances of noninstinctive behavior are of adaptive character such as may better secure survival. One might expect reproductive behavior, this being so fundamental and indispensable a condition of group persistence, to remain unaffected by any evolution of adaptive behavior conducive only to individual survival. However, as was remarked in the preceding chapter, the advance of the more evolved vertebrates toward full domesticity was not always a progress that left their species more assured of persistence, and the ethologist finds that reproductive behavior, too, no less than adaptive behavior, can be quite radically changed as a result of nurtural conditioning during growth.

The nightingale reared in isolation from its kind will at maturity sing any song it happens to have heard, and of late a more attentive research has found that much the same appropriation of specifically alien behaviors occurs in many species. (It has even been reported that certain migrant birds briefly sojourning in England on their way to the mainland of Europe arrive there with a discernible

English bird-accent.) In these avian groups, even the mode of reproductive behavior is directly rather than genetically transmitted, for bird-song is quite an important feature of avian reproduction, its function being to preempt a hereditary breeding-station and to attract a specifically appropriate mate, so that breeding is contained within the species. But much more conclusive is the discovery that birds reared by a foster-parent may at maturity refuse to mate with birds of their own species, and make sexual response only to birds of the fosterspecies, for here we are shown how nurture may wholly erase one specific behavioral mode and put quite another in its place! How thorough, complete, and radically effective may be this direct transmission of reproductive mode to growing progeny is indicated by the fact that birds reared by man may at maturity find sexual stimulation only in their human attendants, their specific avian response having been replaced by a quasihuman response. The ethologist speaks of all such behavioral reconditioning of growing animals as "imprinting." One of the most articulate pioneers of this rapidly developing science has reported the following tragicomic case-histories of humanly imprinted birds:

"My friend Hellman in Rotterdam got a male budgereegah imprinted upon himself. The next year, he imprinted a female of the same species upon himself. You may know, budgereegahs are very good at accepting substitute objects. Normally imprinted budgereegahs will take a celluloid doll for a substitute female, try to feed it and attempt copulation. Now, when Hellman put these two birds together in a garret room and shut them up, they gradually became conditioned to each other and accepted each other as substitute objects, copulated and even reared young . . . Now, the point is that after these budgereegahs had been together until they had reared two broods successfully, and were about to rear their third brood, we exposed them to human beings. I was standing there, and the partition was removed. The budgereegahs saw a human being after a very long time, after having been "married" successfully for a very long time. Both at once went into a frenzy of courting directed at the humans. Then, they had a terrible fight between themselves and let their present brood die of hunger because their whole reaction to each other had been disrupted by the reexposure to humans. The experiment was over and we did not want the young to die, so we retired immediately, but it was too late; this brood of young starved to death.

"J. Portielje, director of the Amsterdam Zoo, had an analogous case in a South American bittern, **Tigrisoma.** This bird was slightly ambivalent in his reactions, still he preferred Portielje to his female. In order to breed them, Portielje put the pair into a ground glass cage in which they could not see any living being except each other. This was necessary because the male at once wanted to court human beings and chased the female out of the nest. I was very impressed when Portielje entered the cage. The humanly imprinted male immediately drove the female off the eggs and invited Portielje, by ceremonies, to sit on the eggs. Portielje is a man a head taller and also much bigger than I. It was very funny to see that wisp of a bird inviting and emotionally urging Portielje to sit on his nest."[9]

Those budgereegahs tell us a good deal about ourselves. They help us to understand the sudden infatuations that have disrupted many a staid middle-aged menage, and they add an explanatory footnote to the plaintive ditty that called for "a girl just like the girl that married dear old Dad." They leave no doubt as to the manner in which even the behavioral modes of reproduction themselves, these being the very constitutions of living groups, are directy inherited, or "acquired," in these highly domestic species. Here is a living nature very different from what we are accustomed to look for in a realm of "animals governed by instinct," or even to envisage very clearly within our own realm of mankind, which still must transmit its own genes and genotypes.

It is quite evident that nurture disturbs, inhibits, and deflects into new directions what would otherwise have been a genetically determined course of behavioral and bodily development, to bring all behavior into a state of fluidity, the combination of individual differences with the particular circumstances of nurtured growth leaving adult behavior largely unpredictable. And it must have been long ago, at least 300 million years ago, that animal nurture thus unlocked the genetic gate that much later, with the achievement of full domesticity, would be swung wide open, to insure that every human response would carry its proper behavioral imprint. Not down from an Olympian sky, but up from the ancient mammalian womb or premammalian egg, came the human behavior that is still a product of postnatal nurture, complementary to long gestation.

Soon, we may hope, a rapidly developing science of ethology will adequately describe the accelerant progress, successively indicated in its earlier and later phases by vertebrate, mammalian, primate, and hominid behavior, that led to human behavior as we know it today. This behavioral history will incidentally describe the steady dissolution of the species process as a result of the withdrawal of early growth from selective pressures, with its replacement by a social process that gives to human populations a steadily enriched "pool" not of diversified genotype but of directly inculcated behavioral resources. For, as a geneticist has written: "Man has acquired what amounts to a totally new evolutionary system . . . conceptual thought and language constitute, in effect, a new way of transmitting information from one generation to the next. This cultural inheritance does the same sort of thing for man that in the subhuman world is done by the genetic system. . . . This means that, besides his biological system, man has a completely new 'genetic system' dependent on cultural transmission."[10]

As we look back upon this behavioral evolution, we come slowly to comprehend its significance as a transition from a protozoic mode of group regeneration to a fully metazoic mode of life. Only man can be said to have completed that transition, every animal group having stopped somewhere short of it. In the lower animals, where all behavior is instinctive, every sort of behavioral readaptation must be transmitted through the reproductive cell—and how narrow is that passage, no wider than the dot over this i! For the somewhat more domestic animals, the channel is enlarged sufficiently to permit of some direct transmission of adaptive character conducive to survival, but further to widen it brings an impairment of reproductive character that threatens life itself. Only in man do we see safely completed the transition to fully metazoic character. Here the whole organism, and not only the living speck that is a reproductive cell, may directly transmit its individual character both in mode and type down the tides of time, to become through the agencies of symbolic communication a shaper of the future.

VIII

Human History: The Small Societies

Being aware of the outcome of the evolution of life in mankind, we can find in protozoic history, in invertebrate history, and in vertebrate history a record of three successive stages of the progress to man. There follows next human history, but man is evidently still in the making. How may one describe human progress in ignorance of its outcome?

Man is God's image, the only creature possessed of a soul. He is **Homo sapiens,** the species endowed with reason. He is the fire-maker, the tool-shaper, the weapon-wielder, the supremely aggressive animal. Every definition would require a different interpretation of human history. We, however, are required by the foregoing description of animal history to define man as **Homo domesticus,** a group differentiated from all else that lives by its fully domestic mode of reproduction. No animal group is thus raised above the selective pressures that bear upon other life.

There is psychological and perhaps other evidence to support the hypothesis that man arrived at human estate by way of a rapid change, conceivably the result of some gene mutation, which so retarded maturation as to more than double the duration of human growth, and possibly add considerably to human stature. Contemporary psychology finds a strange ambivalence in man's behavioral development. It is as if there were superimposed upon an infantile growth, one that would bring maturity at five or six years of age, a second stage of maturation. The precocious sexual development is inhibited and suppressed, so as to allow of domestic conditioning through long childhood to delayed and difficult adolescence. One almost thinks to discern here the traumatic scar and behavioral seam that were left by that last deliverance from an animal past.

The group progenitor to mankind, as it gradually divested itself of the specific character that had kept its populations closely adapted to their habitats, did not become just a structureless plurality of individuals. The parental care of several progeny of different ages gave rise to the family, and the nurtured progeny brought into their adult relationships the behavioral habits inculcated by family life, so that what ceased to be a species became a society. The direction of human progress is consequently to be found only in man's social evolution.

Anthropologists are inclined to date the beginnings of social history quite far back in time. One of them writes: "Life in structured social groups is characteristic of primates. What we find as the common social core of all but the lowest primate groups is the continuous association of both sexes with their offspring."[11]

Primate history would cover a period of some fifty or sixty million years. It is true that for simian and hominoid groups reproduction is a process very different from the brief seasonal conjugation of the less evolved animals, and one that increasingly becomes the large preoccupation of adult life. Yet we may not find there the beginnings of the social evolution of man. The reason is that the primates have remained subject to the selective pressures borne upon species by species and aggravated by animal increase, whereas human populations, liberated from those animal pressures, are subjected by human increase to quite another sort of pressures, so that human history is like no other history.

However, there is one broad feature of the evolution of animal life that reappears in the evolution of societies, and it may offer a convenient approach to human history. This is the tendency of living groups of divergent descent to converge toward very similar sorts of external adaptation and visible appearance. Convergence is usually found wherever a later expanding group displaces an earlier expanded group, occupies its habitats, and then proceeds to an evolution of adaptive characters quite similar to those of the group it has replaced.

The most familiar example of convergence is probably that between the mammals and the marsupials. The mammals displaced marsupials everywhere except in Australia, where marsupials remained much what they had been. Elsewhere, the mammals proceeded to an evolution of adaptive habits so similar to those of the marsupials they had displaced that they often came to resemble them closely in their bodily appearance, and Australian settlers, as they encountered marsupial species, could name them for more familiar sorts of mammals. Much the same sort of convergence is found by paleontologists between the mammals of today and the extinct reptiles they long ago replaced.

Once alerted to it, the eye finds convergence everywhere. Bats pursue insects at dusk much as do swifts and swallows. The roadrunner courses his prey like a fox or coyote. The aquatic bird forages under water like any fish—to look down upon penguins swimming underwater is to think one sees a school of rays. The staccato go-and-stop flight of the hummingbird resembles that of a dragonfly or wasp. Adaptive type rests lightly upon the deeper seated reproductive mode, so that similar modes may support very different types, or different modes support quite similar types.

Much broader is the convergence that appears in the parallel evolutions of species and societies. Although human life proceeds upon a biotic level raised above animal life, human populations still depend for their persistence upon the reproductive processes of their shortlived individual constituents, and societies no less than species are given identity, integrity, and longer or shorter persistence by their own internal regenerative processes. Just as are species, so societies are differentiated by the particular modes of regeneration they transmit down their ephemeral generations. From the time they began their expansive radiation, societies have exerted upon each other pressures that were steadily intensified by human increase, and it was to their modal behaviors, no longer genically transmitted in instincts but directly inculcated by postnatal nurture, that they owed their social identity, their behavioral adaptability to internal stress and external change, and their consequent duration.

The early societies were necessarily small and scattered. Not only did a completely primitive economy support but few individuals upon considerable terrain, but the societies, being still without speech, depended wholly for their group-integrity and persistence upon their direct nurtural transmission of all behavioral adaptation, of mode as of type. Even late primitive societies persisting alongside of civilization, these societies being equipped with fire, speech, housing, and many a useful artifact, numbered only two or three hundred individuals, and anthropologists believe that the earliest societies would contain not more than forty or fifty. Integrated by parental nurture alone, they persisted in virtue of the strong quasi-domestic loyalty that binds a small and inbred clan. Only when human increase brought congestion would there be generated those new pressures, exerted by societies on societies and so much stronger than any imposed upon species by species, that were to usher man out of Eden, and give to human life the not untragic character it has since retained.

The small societies, we can see, would require three sorts of adaptation in order to persist, and, as these were behavior-patterns directly transmitted down the generations much as they are today, nothing prohibits our use of current nomenclature in their description. Indispensable for day-to-day survival was the economic adaptive type that fitted the small group to its temporary habitat, and this was imitatively acquired, the children observing how their elders dug roots and gathered fruits each in its season, and knew where and when and how to snare fish and flesh and fowl. Second and of intermittent but increasing importance was adaptation of political type, similarly transmitted, enabling the society to defend its terrain in face of encroaching neighbors, as human increase brought congestion with its stronger pressures. But third and all-important was the faithful transmission of the behavioral mode of group-regeneration, or nurture, which provided the channel through which all adaptive character, too, was now propagated. And here we are shown the three factors constitutive of social process and determinative in their diverse modifications and ceaseless interplay of the social change that is human history.

In an earlier chapter mention was made of the power of bisexual reproduction to secure an individual differentiation of progeny. Most crucial here is the process of meiotic reduction, or meiosis, which gleans from two parental assortments of genotype the single assortment that is bequeathed to a new individual. It is to meiosis that a living group ultimately owes the continual diversification of individual character that keeps it adaptable to its own and other change.

However, in mankind, sexual conjugation is only an initial phase of the long process of domestic reproduction, or nurture, which properly covers the whole lengthy period of human maturation, and it is by way of nurture that human parents transmit to progeny, not genetically but directly during growth, their own behavioral characters. We must accordingly recognize in human development a distinctive process of **domestic meiosis,** the generative function of which is to bring to each child, already given a unique combination of parental genotypes, also a unique combination of parental behavioral traits.

Humans, in short, are twice individuated, first genetically and then nurturally, and this process of domestic meiosis with its behavioral synthesis may continue long after bodily growth is seemingly completed, individuals who had earlier behaviorally "taken after" one parent sometimes coming later more to resemble the other parent. Human growth is a protracted, usually difficult, often precarious business, as is shown by the incidence of neuroses which indicate an inability to resolve incompatible bequests of paternal and maternal character, and which may more ultimately be due to conjugal incompatibility between the parents, whose own behavioral differences must be accommodated by good conjugal association before they are domestically adequate and able to bring proper nurture to the progeny. As the depth-psychologist now discovers and reports, this domestic meiosis is the very process of human growth—for each of us humans it is the life-process itself, which when it is successful generates the human character we call "personality" and properly esteem. And it is given its impulsion and direction not by sexual passion but by conjugal devotion, filial piety, and parental love. Not by nine months of pregnancy, but by a dozen or score years of upbringing, is man gestated.

Because the several children of a conjugate pair are at birth already genetically individuated, they instinctively respond to parental stimuli with a difference, which then inevitably elicits from each parent a somewhat different response, and the result of these innate individual differences is to make each domestic affection unique, so that it develops as one facet of the growing and diversified association that is a human family, this being a new unit of planetary and cosmic life arisen out of the earth with man. The growing child is instinctively aware that in its unique personal relationship with the parents, a tie of which it is only lethally to be deprived, lies its vital umbilical security, its integral identity and self-esteem, even its whole prospect and hope of life.

Through the parental, filial, and fraternal affections of the home there develops through a dozen or score of years the living unit that is a family. Through this domestic association the children may be given, without any infraction of their personal identities, a common and recognizable behavioral mode which, when they mature and raise families of their own, they will substantially transmit to these families again, for their choice of a mate and their every domestic disposition will be deeply conditioned by the domestic character given to them by their own nurtured growth, so that they propagate in new families a behavioral mode similar to that which they had themselves acquired.

It is in this way, each generation of ephemeral families producing its successor and transmitting to this successor behavioral modes newly diversified by domestic meiosis, that societies are regenerated and better or worse preserved. Not being species, societies are not regenerated as are species, but they are domestically regenerated. In this realm of human nature it is families that reproduce their domestic character in kind, to make it the mode or cultural norm of their societies. Ethologists are increasingly inclined to recognize in the animal family an effective living unit, in terms of which any evolutionary change of the higher animals may have to be described, but it is presumably only in human history that is displayed a fully domestic and social evolution, one that is in no degree confined by the restrictions imposed upon every animal group.

It is in the small early societies that we might have expected to find these group-processes most clearly at work. As human increase brought congestion, and subjected the societies to strengthened selective pressures, these being always political pressures exerted by societies upon societies, successive generations would be compelled to devote more of their time, thought, energy, and material resources to meeting the pressures. This is again our civilized condition, and the pressures would bear more immediately and palpably upon those primitive groups, which existed from hand to mouth and had no surplus of anything. But whatever is given to adaptive type may be withdrawn from reproductive mode, and in human life all such preoccupation with what is of immediate survival-value may be bought at the expense of domestic behavior, adequate nurture, and that provision for the human future which alone keeps a society, whether this be primitive or civilized, adaptable first to its own internal change and thereby adaptable to external change, so that it may still persist.

As human increase left the societies congested, there had to be more forceful discipline providing a militant defense of terrain, feeding-grounds, and material resources. With strengthening pressures the group-character of the society underwent a deteriorative change, parents giving less time and thought to nurture, and children being inducted into adult services at an ever earlier age, before they had acquired the humanly domestic character upon which ultimately depends every transmission of behavioral mode and type. So, in place of authentically human growth there came a precocious and too extrovert growth, and in place of the domestic sociability of the healthy society there came the uniform regimentation that leaves a people in appearance integral and intact, but under its rigid solidarity and outward discipline increasingly brittle and dissociate. This internal desiccation rendered successive generations progressively less capable of transmitting a domestic character they had never themselves fully acquired, so that even when external conditions became less stringent this alleviation need not bring recovery. In the excessively militant society the family itself can be effectually dissolved, and the very process of regeneration then becomes an external political adaptation that provides manpower needed to replace casualties, intimidate neighbors, and enlarge terrain. With sustained human increase there came a deterioration of domestic life and impoverishment of communal resources that formal piety and exacted loyalty could only disguise and deepen, and a society that was really impaired in its domestic process and human integrity had but small prospect of betterment.

As we were shown by our study of evolutionary change, it was never the most vigorous of living groups, on the contrary it was species left less viable by their too uniform character that were prone to expansive increase. And it would now be societies that had made some exchange of human nurture for efficient political adaptation that enlarged their domains, to know rapid increase and to be disruptively multiplied into congeries of closely cognate societies. Inevitably, following upon such inordinate expansion, would come accelerant decline, with a later displacement of these degenerate populations when other societies, better integrated and more adaptable, but earlier kept obscure by their geographical isolation or through some wise disinclination to barter domestic health for political power, were eventually brought by new and stronger pressures to move to militant defense, only then to be carried in their turn through expansion to decline.

Early human history may always remain conjectural for lack of evidence of its passage. The archeologist can record an advance to better stone implements during the last several hundred thousand years, but he learns little about the groups that fashioned and used these tools, and the rapid expansion of successive civilizations during the last few millennia has blurred the ethnic distinctions that might have disclosed the genealogy of early man, somewhat as the classification of species points to the genealogy of plant and animal life. However, what the anthropologist does vaguely discern through his synthesis of archeological, ethnological, and philological evidence would tell of expanding and migrating peoples that crossed whole continents in rapid succession, with each new wave driving before it or absorbing into itself the remnants of what had earlier been as widely dispersed, and we appear to be directed to an evolutionary pattern not unlike that displayed by the animal groups in the evolution of species, except that the great expansions and replacements which covered epochs and periods of geological change for animals are to be measured in thousands or tens of thousands of years for humans.

This convergence or parallelism between social and animal evolution really describes a process of devolution characteristic of both realms, but human populations nonetheless differ most radically from animal populations in one all-important respect. However far societies may be expansively radiated, and however wide may be their cultural divergence, they remain indefinitely fertile with each other, whereas animal populations are multiplied into completely separate and intersterile species.

The instinctive behavior that binds an animal within its species leaves it innocent of malice in its often lethal response to what is not its species, but in man this same response, still genically determined, leaves the individual instinctively associate with all that is human and always intelligent of his kind. The pressures exerted by societies upon societies have always of necessity done violence to the constitutive nature of man, even as they do today.

The direction of human progress was necessarily one that increasingly brought to expression and effective power that deepest association of mankind, so that it might moderate, assuage, and eventually bring under control the destructive forces generated by human increase. There was required a mode of regenerative behavior that would keep the small societies together yet apart, disparate in their particular cultures yet still effectively associate, somewhat as local populations are integrated into a durable species. This requirement was met by an advance to tribal organization, which was bridged upon a new mode of symbolic intercommunication.

It was to implement that largest and deepest instinct that speech came. The beginnings of sonic communication are found in the birds, the reptiles, and even the insects, and the evolution of language is accordingly the outcome of a behavioral progress sustained through several hundred million years.

The forms of symbolic communication found in animal groups are usually implemental to conjugation, and serve to secure breeding within the species. The nearest known approach to human speech may be found in bird-song, a reproductive adaptation which in its variations upon a specific melodic theme resembles a language which maintains throughout its many particular utterances an identical sentence-structure, or syntax. The most evolved forms of instinctive symbolic communication are exercised in the courting behaviors which stimulate and contain reproduction within the breeding-group, and we find that among some birds the reproductive behaviors have been elaborated into the group-rituals which accompany and direct seasonal breeding. We may reasonably infer that these rituals, among which is the beautiful rite of the dancing cranes, evolved when lengthening nurture began to dissolve the species-process, so that instinctive behavior no longer sufficed to confine interbreeding within the species, for the rituals are clearly a behavioral elaboration of courting ceremonies which earlier were genetically transmitted.

The group that was to be multiplied into mankind must have replaced most of its genetically transmitted instincts by directly transmitted behavior-patterns long before it became a fully evolved human society, and we may believe that human populations were never without those rites of spring and fall which brought them to periodic congregation, now that they were no longer genetically compelled to seasonal reproduction. Out of those communal exercises auxiliary to social regeneration evolved the totemic rites, which symbolized the particular society by way of some selected plant or animal species, the ceremonial behavior being built about a branch or sheaf or flower of the totemic plant or the hide or head or horns of the totemic animal. Here we find the symbolism that antedated speech and could lead the way to speech!

For, it is to be observed, this use of plant and animal kinds to designate particular societies differentiated itself from earlier symbolic behavior by its recourse to mimicry, to which the totemic cult now gave quite a new importance. Mimicry is behavioral description, achieved without use of language and making of the whole organism an instrument of symbolic communication (and in the histrionic art of pantomime mimicry remains a most direct, effective, and universal medium of dramatic expression). Through its use of mimicry the totemic rite effected a synthesis of adaptive with reproductive behavior that was of quite inestimable consequence, in that it has given to all later social evolution its distinctively intellectual character.

Humble indeed was this mimetic origin of the human intellect. An animal inherits with its genotype an instinctive intelligence of the several species upon which it feeds or of which it may be the quarry, and, when this built-in knowledge was exchanged for better nurture and lost to man, it had to be given its behavioral surrogate. Still instinctive in the well-nurtured group, however, is the play-impulse by which nurtured progeny elicit associative response from parents and siblings. And how might human progeny be more easily or effectively behaviorally reconditioned, or "imprinted," than through directed play, with the adult displaying in behavioral pantomime what the child would later encounter in its adult environment? So the human adult identified himself mimetically now with this animal and now with that, mimicking the specific call and characteristic motion of the animal, and doubtless evoking from the child a half-fearful yet delighted response, with the child persuaded to pursue the "prey" that after appropriate and instructive evasion and struggle would duly deliver itself up to its small human superior.

No child's play but a desperate recourse of adult concern was the mimetic symbolism of the totemic rite, which can never have been the merely adaptive exercise, preparatory to hunt or skirmish, as which it is too often discounted. The primitive rite would exact a total dedication and dispense a religious consecration, for there is nothing in religious or artistic symbolism that does not have its lineal descent from those communal exercises of early man. Its ritualistic purpose and vital function was to generate the group-integrity that alone brings social health and group-persistence.

By their exact synchronization of vocal and motor expression the participants in the totemic rite were brought into a physical, physiological, and psychological unit that was possessed of dynamic actuality and exerted a palpable power; for the individuals experienced a lapse of personal identity due to their "possession" by another larger life that both was and was not their own, and those who emerged from the rite were not just those that went into it. We may perhaps speak of the **social meiosis** by way of which the participants in a totemic rite were bodily translated into the larger and more enduring life that was their society.

As the millennia followed one another, and each new generation religiously re-enacted the guarded rite and exactly transmitted to the next its behavioral pattern, with chant and dance mimicking the call and motion of the totemic beast or simulating the sway and rustle of the totemic plant, certain sounds came to be nouns that named the totemic property with the repeated exhibition of which the sounds were exactly synchronized, while other sounds became verbs which symbolized the mimetic movements, and the single rhythmic cadence of both chant and dance provided the scansion that would become the sentential syntax of a nascent language.

The vital power that worked in the primitive rite is made most manifest today by the tenacity with which the totemic symbols have retained their hold upon the mind even of civilized man. Long after the totemic symbolism had been outmoded and replaced, the magic incantation of the chanted word would shed its blessing or hurl its curse. To this day the totemic eagle soars, the lion roars, the cockerel crows, the bear walks more than ever like a man, the dragon wakes from sleep, and in a hundred thousand dim chapels and candle-starred cathedrals the branches of the true vine, mixing their totemic metaphors, drink the atoning blood of the totemic lamb.

Are there not apparent in this totemic symbolism the first stirrings of the human intellect, this being a faculty quite uniquely differentiated by its behavioral response to nature at large. The plant and animal totems made articulate an appreciation of biotic patterns and an intelligent conspectus of natural structure, later to be elaborated in a classification of species. To primitive man, as to Aristotle much later, the specific forms of plant and animal life must have seemed to be immutable and ideally secure, so that they displayed to man the eternal structure of nature and the true substance of being. Just so changeless would he have his societies be!

Even as the specificity of nature gives to other life its assured persistence and changeless design, he opined, so must it give persistence and immutability to a socially patterned mankind. Because the totemic symbolism related the pattern of human life to nature at large in this articulate way, it inaugurated an intellectual quest pursued not just for its own sake, as we used to say, but because it promised a revelation to man of his true identity, with an understanding of his due place and rightful function in the cosmos.

Human History: Civilization

The shift to civilized economy came as a direct consequence of an advance to large-scale cultivation, which in its turn was the result of an earlier domestication of plant and animal kinds. As the latter does in fact bring to nonhuman growth something like nurture, it is perhaps not unaptly named. It may have begun simply with the observation that seeds flung upon a midden had sprouted to bear new fruit, but the persistence down to our own time of primitive taboos forbidding the use of what is eminently edible but ethnically "unclean" suggests that it arose as a by-product of totemism, certain plants and animals held for ritualistic use later coming to be bred.

To begin with, the cultivation of crops would only supplement a still primitive economy, but, in desert oases where increasing desiccation restricted vegetable growth, recourse was had to a rudimentary irrigation. It was when this adaptive innovation was brought to the great river deltas that there developed the large-scale cultivation supporting civilized economy.

Plato believed that civilization originated in Egypt, and the Egyptian hieroglyphics with their many animal symbols do seem to be the most primitive script. However, the wet soil of the Nile delta defeats archeological excavation, and it was in Mesopotamia (that arid "land between the rivers") that the inquirer could dig down to the ruins of primitive villages upon which the earliest cities were erected, so that whole histories of civilized expansion and decline were given their legible record. The oldest of these cities was Sumer, built at the head of the Persian Gulf by settlers believed to have come there by way of the sea.

The newcomers who built Sumer could have had no foreknowledge of the great cities, congested populations, and unstable empires that were to follow from their enterprise. Experienced irrigators they evidently were, able to see that the great swamps of the delta, once these were drained and cleared, might be made an arable paradise. Resolutely they went to work, which once begun had no stopping. Year-long sunshine, inexhaustible water, and deep alluvial silt deposited by the great rivers produced unprecedented harvests, and humans had probably never been so securely and abundantly sustained. There was naturally unprecedented population-increase, too, but more swampland had only to be drained and cleared, sown and harvested, to fill any number of mouths. Where primitive economy had supported hundreds at most, this civilized economy sustained thousands, and it came to support tens and hundreds of thousands, finally millions.

A good deal more than some small skill in irrigation, however, was needed to effect the momentous transition to civilized economy. Just as does any living group, those Sumerians carried their vital resources wholly within themselves, and between them and primitive man lay millennia of social evolution, which had brought the small early societies into wider tribal association.

Primitive man, dependent as he was for his whole sustenance upon what nature directly provided, had considerable intelligence of plant and animal life. Perspicacious was his discernment of the congruence between species and societies. This had supported his totemic culture, and it probably also conveyed an awareness of the symbiotic relationships that weave plant and animal kinds into a self-sustaining ecological system. So, he may well have believed, that a human ecology could extend to each and every people an assured persistence within the realm of mankind. At the least, the totemic symbolism must have served to keep cognate societies aware of their common descent, and thus opened the way to wider tribal association.

On the other hand, any advance of this sort would eventually lead to human increase, and thereby intensify pressures exerted by peoples upon one another, eventually to lead to intertribal conflict. It is evident that early man came to find the wisdom drawn from a survey of plant and animal species still to fall short of his need, and he accordingly raised his intellectual sights to seek and seemingly find a more potent magic in the sky.

Out there in the eternal stars, he could believe, is visibly displayed the Great Wheel that by its impeccably uniform motion generates a cosmic life. In the nightly revolution of the majestic constellations the primitive seer now thought to find the redemptive measure that must restore to human existence its proper stability, and bring release from the vicissitudes of social change. The exuberant chant and vigorous dance of the totemic rite were exchanged for the stately procession and sonorous polyphonies of a celestial cult, one that would find in nothing less than cosmos itself the foil and proper counterpart of man. (Yet the ancient totems would not be wholly displaced and forgotten. Pisces the Fish, Aries the Ram, Leo the Lion, and other signs of the Zodiac remain to remind us of the lineal descent of our exact science from an older and more vitalistic faith.)

No dynamic process of cosmic generation, but only an illusory optical effect of the spin of this earth upon its axis, we now are aware, is that impressive procession of constellations across the night sky. Both truth and error invested that celestial myth. Out of its ingenuous astrology would be delivered a classical science that was eventually to escape from its geocentric predicament yet still cleave to its deeper error, which was to discern in all natural occurrence only an eternal uniformity, or "universal law," that must forever hold all that is in thrall.

There are temple ruins much older than is civilization. Long before urban civilizations appeared, the sacred site dedicated to the celestial cult and prescribing its ceremonial procession had been given a third dimension. There now arose a shrine which through its exact orientation and cosmically ordained proportions should even more effectively bring into human life the divine attunement that gives to all that is an eternal composure. We can vicariously share the ritualistic devotion of an Egyptian ruler who uttered the following incantation as he stretched the cord that defined the site of a new temple:

> "I have grasped the peg with the handle of the hammer.
> I took the measuring line with the goddess Safekhabui.
> I watched the advancing motion of the stars.
> My eye was fixed on the Bear.
> I reckon the time checking the hour,
> and determine the edges of thy temple.
> I turn my face to the course of the stars.
> I direct my eyes upon the constellation of the Bear.
> There stands the time pointed with the hour.
> I determine the edges of thy temple."[12]

That priest and king was probably no astronomer, and only religiously intoned what he had learned by rote, but many millennia of astronomical observation, mathematical invention, and architectural experimentation had devised the formula that would so long be mechanically but piously applied. Exact science far antedates civilization, and it long had its laboratory in templar architecture, so that astronomy and geometry would remain "the sacred sciences," and imprint upon scientific thought the quasiarchitectural character it still too much retains. Geometry was not created by the Greeks from whom we had it, nor was it ever just the rule o'thumb of Egyptian peasants replacing landmarks swept away by the Nile's annual inundation. Never just the child of economic necessity, this remarkable discipline came to civilization with those who first built civilization, and they had it from earlier temple-builders, to whom it came from the creators of a celestial symbolism in a remoter past, when the only shrines to be erected were the monoliths that contained the ceremonial procession, and geometry was most literally an earth-measurement.

The governing motivation of the builders of Sumer was of the sort we would call religious, which is to say that it was authentically social and regenerative, not just political or economic. The scientific technology that enabled those pioneers to bring under productive control the floodwaters of the great rivers was no rudimentary knowledge. It was a by-product of millennia of temple building intended to make of human society a behavioral microcosm processed even as is cosmos itself. It was the conjunction of this intellectual social culture with the external conditions provided by the delta, not just those conditions in and of themselves, that built civilization.

Earliest Sumer had no polity! The temple was spoken of as "the household of the gods," and it was in fact the whole social economy. For the temple owned and was all buildings and land, crops and herds, the laborer in his ditch and the priest in his precinct. The methodical erection of the great shrine, the draining and clearing of swampland, the sowing and cultivation and harvesting of crops, the making and care of tools, the storage and just distribution of produce—all were elements and phases of a single sustained ceremonial exercise, each item of which was exactly prescribed and religiously executed. Daily, even hourly, a templar rite evoked the spell that kept a people synchronized in all of its corporate activity, quickly and efficiently responsive to a direction that was socially responsible and technologically equal to its task. Never out of any merely adaptive behavior fitting an individual to its habitat, but out of the courting ceremonies and group-rituals that had once secured only intraspecific breeding had come with the fullness of time, by way of totemic and celestial symbolisms, this templar economy that was now to become civilization.

Given this happy conjunction of social resources and material conditions, the colony prospered and proceeded to expansive increase, and, just as naturally and inevitably, the populations so enormously multiplied became unduly distended, and rapidly lost their domestic health and social integrity. Sumer was amply provided with sunshine, water, and alluvial mud—but with nothing else, it had to import all of its metal, wood, and even stone. Around the temple precincts there came to be great markets to which flocked traders from far and near, and there arose that babel of alien tongues which, it was later alleged, finally brought the whole templar economy to behavioral confusion and material collapse. We may guess that the tale of the Tower of Babel perpetuated the parable of some anonymous and forgotten sage, one whose clearer vision had perceptively observed a people become socially incommunicant and at its heart disrupt, now that the market place had done its secularizing work, and set a city between a laity lapsed into peasantry and a priesthood become bureaucratic and esoteric, which is to say under all of its verbal piety religiously apostate.

But there the record stands! Upon a hundred feet of excavated clay we are told how social concern hardened into political power imposing its physical compulsion by militant might, how priests became kings and tithes became taxes, how traders became merchants and merchants became rich and bought land to become lords, how men became peasants and souls became serfs. Broken was the pitcher that had contained a fluent and self-regenerate life, loosened was the silver cord whose sweet attunement had kept a people communicant and one, the melody that had molded action and quickened thought was stilled.

Further settlements were made upriver, as continual increase brought congestion not to be contained, and there came other peoples to copy the new economy and to raise new temples about which grew new cities. And still the cities grew, so that in place of a single shrine there arose a cluster of temples housing as many deified totems, and "the household of the gods" was no more. Similar in their civic institutions, which is to say in all of their external adaptation, political and economic, those earliest city-states had little to differentiate their now secular cultures, no memory of a social past, no prospect upon a historical future, neither place nor function in the living dimension of time that is creation; and they were inevitably carried through the fatal course made familiar to ourselves by Greek history, as each city was expanded only the more zealously and belligerently to compete for water, land, and commerce. Their destructive feuding finally made it apparent that only imperial control could conserve and allocate the waters upon which all alike depended for sustenance and survival, and now arose Sargon, first of the long roster of world-conquering great, hailed doubtless as a savior from priestly misrule, and followed by a procession of potentates who founded short-lived dynasties figures-heading ephemeral empires.

In pre-Columbian America, we are told, "interest in permanent public or ceremonial building is a correlate of settled life, it does not appear until after sedentism is well established," and we know that sedentism could not become the habit of life until efficient cultivation had obviated the need of continual migration from soil become exhausted to new arable land. The interest in ceremonial architecture, we are further told, then "undergoes changes suggesting a drift away from the temple, to the palace compound, and from this to military construction and urban planning . . . The evolution of the city and the state in native America appears to be linked with the growing power of the military and secular forces and with attempts to propagate religious, political, and social views."[13]

Imperial Mesopotamia collapsed and was replaced not once but many times, its successive capital cities giving to it as many different names. But restoration never brought health, it only postponed extinction. Harsher and more exacting became the laws, more preposterous the presumption of the self-deified rulers, but social cohesion could not be compelled, cooperative communion was not legally to be enjoined. Still the canals choked and the dykes crumbled, until the stupendous irrigation works fell into irreparable disrepair, and, as ever more of the once fertile fields lay fallow, the waiting wilderness silently reclaimed its own. There would come a day when the airborne traveller crossing these Mesopotamian wastes would unbelieving look down upon the fossil outlines of great but nameless cities, buried so long under desert sands that no last memory of them remained.

We look back upon ancient Mesopotamia with some wonder, astonished and not a little dismayed that so imposing a civilization should so completely perish from the earth, even what it bequeathed to later posterity bearing always some other name upon its title page. But many another civilization has just as completely vanished from sight, and was that history intrinsically remarkable, or does it impress us because its last days were biblically numbered? Contemporary archeology uncovers a most legible evidence of its passage—much as the ammonite shellfishes through their well-preserved fossils make spectacularly vivid a wholesale extinction and replacement of species that was and is the rule of plant and animal nature.

Mesopotamian history provides a suitable introduction to several thousand years of civilized history. We are repeatedly shown how primitive peoples disturbed by great civilized neighbors have themselves in turn been carried, as a result of their appropriation of civilized adaptive habit, political and economic, through rapid and wide expansion to as rapid decline.

First adopted, the better to meet civilized pressures, are the weapons and militant habits of civilization, with an expedient and provisional organization of tribal units into a loose federation, but even this quasitribal association conditions economic expansion which soon calls for stronger political support, and in internecine conflict the feudal powers are exhausted and dissolved to make way for national polity. Customs are codified, new laws are enacted and administered, the economy prospers, and the monarch soon finds himself no longer a feudal chief but the symbol and agent of a rising mercantile class. There comes now a happy phase of national growth, with much appropriation of civilized culture in a "renaissance" which actually uses literary and artistic symbolism to bring to expression the emergent character of a young and expanding nation. And here the new civilization displays its short bloom, for with sustained increase the economic needs conditioning survival come to assume dominance over all other interests, and the rapidly expanding people soon forgets even its short national past in its diligent and ever more efficient provision of the economic conditions of day-to-day existence. But this relase from all social commitment is taken to be emancipation, and the distended civilization now becomes consciously "modern" in its formal rejection of whatever does not comport with its fully civilized mode of being. It increasingly identifies actuality with economic fact, power with possession, health with individual well-being, wisdom with know-how. Human life loses its **raison d'etre,** and what had been a pursuit of natural knowledge becomes a technology which further expands industry, augments production, and accelerates increase, while there is ceaselessly elaborated the code of civic law that supposedly gives to a people stability and persistence, even as other nature is supposedly preserved by a "natural law" which in truth points only to a uniformity of character that is the sign and attest of nature's universal dissolution.

Was it those arresting words **Decline and Fall** that gave to Gibbon's classical work its immediate notice and lasting claim? They toll a knell of vanished pasts that brings to every civilized breath a premonitary constriction, to every civilized pulse a suspended beat, as if a sudden shift of stance had momentarily revealed that waiting death. To dwell too long upon this abstract concept of civilized devolution is no more to see the rising civilization still unborne upon its primitive heritage, it is to envisage the broad full-fledged civilization, widely expanded, consciously modern and mature, and already sophistically bent to its decline. For even now is evident the vacuous eye and inert complacency with which the distended civilization mirrors in itself an eternal cosmic design, and, knowing all earlier civilizations to have been mortal, still must deem itself immortal, wrapped and rapt as it is in a timeless present that has no past or future. Shorn of historical perspective and emptied of causal intelligence, it cannot even if it would conceive of its own demise, yet no more can it conceive of why it should endure, and so its turns at last in its aimlessness and ennui and self-distaste to applaud the novelty that rises to displace it, and to connive with secret satisfaction as its own surcease.

We should pay closer attention to the way in which human history recalls the animal history of expansion followed by decline, in order to perceive how one and the same natural process of devolutionary change may work at different levels and within different materials to produce quite similar results. It is a genetic process that brings animal groups to close uniformity making for expansive increase, only to become little adaptable to changing conditions. The vital resources of human populations lie not in heritable genotypes but in directly transmitted behavioral patterns, and in civilized life the disintegrative forces are not imposed from without but are generated from within the civilization itself. It is the dying empire that is given its coup de grace by external enemies, it is only the dead civilization that is overrun by barbaric hordes. Very quickly, in just a few centuries, it would seem, were the social and behavioral resources accumulated through one or more million years of earlier human history dispersed, dissolved, and seemingly lost in the wide and rapid expansion of successive civilizations.

Arnold Toynbee is one of the few historians who have made effective use in the description of human history of the biologist's description of evolutionary change. Approaching his study from a preliminary classification of about a score of extinct, fixated, and still active civilizations, Toynbee found that all of these exhibit much the same sort of devolutionary change. After a civilization has spent the creative impulse that brought it into being, it is disrupted by internal schism into a dominant minority which forcibly imposes its political and military rule upon an alienated and recalcitrant majority. This disfranchised and now dissociate people then makes common cause with peripheral peoples in the establishment of a "universal church." This propagates a new and wider association which gradually supersedes imperial power, and provides a creative impulse to new civilizations which thereafter proceed in their turn to devolutionary decline. The affiliation of later to earlier civilizations through such religious associations gives to civilized history its causal continuity.[14]

The primary concern of modern historians has been a steadily amplified and more accurate chronicle of political events. During this century, increasing attention has been paid to the economic determinants of political change, but only occasional and marginal notice has been taken of the religious, scientific, literary, and other development that have most deeply conditioned the course of modern history. Toynbee's thoughtful and fully documented account of some six thousand years of civilized rise and fall should inaugurate a more scientific historiography, one that will look below the surface of political and economic change—this being at the most only readaptive change—to the underlying associative change that gives to human history its causal continuity with a half-billion years of animal history, so that human progress can be understood as a continuation of the long progress to man. The proper objective of natural science can be nothing less than a description of the natural and cosmic evolution that had its outcome in what exists today.

Toynbee attributed the rise of a civilization to a creative impulse given to a society by its response to some challenge presented by the physical environment. The earliest civilizations arose as a response to the great rivers that could irrigate and make fertile their wide valleys. Greek civilization was a response to the sea that at once separated and conjoined the insular or coastal Greek communities. American civilization mastered the bleak climate and inhospitable terrain that met its colonial pioneers. In fact, however, the transition from more primitive economy to the large-scale cultivation that has supported urban civilization was just an external readaptation to the physical environment, and the regenerative association that initially sustained an expanding civilization was always a bequest from an earlier precivilized people. As Toynbee's description of later civilizations makes clear, these were engendered by religious associations which drew their regenerative powers from still primitive cultures, not from the dissociate and defunct civilizations they superseded. Was not our own politically oriented history a modification of the ecclesiastical history that had chronicled a thousand years of medieval life, during which there was propagated the religious association that would bring to Western civilization its "creative impulse" and regenerative power?

However, Toynbee's discerning and comprehensive description of several millennia of civilized history seems to bear little relationship to what faces us today. Just three centuries of modern history, marked by unprecedentedly rapid political and economic expansion and by an unparalleled progress of industrial and professional skill, have sufficed to bring into question the future persistence of human life and indeed of all planetary life. At any moment may erupt the nuclear Armageddon that could annihilate mankind, and, while explosive human increase demands ever larger sustenance of every sort, material resources are recklessly squandered, arable land is sacrificed to urban development, and the air and water upon which all life depends are polluted and poisoned by industrial waste. Less obvious but more insidiously destructive is the deterioration of every mode of human association, for this corrupts the media of human communication, it makes even our educational institutions the propagators of social demoralization, until at last the family, the very source of man's regenerative power, is left dissociate, and the umbilical cord that gives to a human generation its purchase upon the future is cut through.

This extremity of human need we have quickly learned to put out of mind, because to dwell upon it is to make all our doing pointless. Why beget and rear children, why seek to better our condition, why bother even to reach adult maturity if this earth is but a stage upon which the final curtain falls? Yet should we not ask, before we turn our eyes from it, what brought us to this pass—perhaps to discover just where and when life's enterprise went so far astray?

Even to begin this inquiry is shortly to perceive that our present condition is a logical outcome and either a direct or an indirect consequence of the last few centuries of scientific progress. To this so successful and in itself so admirable pursuit of natural knowledge we owe the technological innovations which in industry and commerce, in medicine and hygiene, in transportation and communication, in weaponry and warfare, indeed in virtually every field of political and economic activity so vastly empowered modern civilization that its hectic expansion came to cover the surface of the earth, what was not bodily included within one or another of its distended empires becoming indefinitely subject to their dominion. No longer are there "peripheral peoples" whose regenerative powers, dutifully preserved from a more primitive past, might revitalize old and engender new civilizations.

Nor, even if there still were such peoples, could we expect to find in their cultural resources the germinal seed of a universal faith that might restore regenerative power to a dissociate mankind. The descriptive science that has so mightily expanded civilization has worked no less effectively to discredit, antiquate, and make unintelligible every faith that was not its own, to leave itself the only faith to which societies everywhere not only can but must subscribe. The actuality it so accurately but still too superficially depicts has no place for gods or other celestial divinities, it provides no evidence of divine law or other supernatural governance of nature, it knows no heavenly providence that can sway the course of natural necessity. The scientist does still verbally subscribe to the doctrine of natural law, but this legalistic language is only a convenient contraction of the assumption that natural occurrence will always and everywhere present observable uniformities of some sort, or at least of some relationship between earlier and later events. Indeed, because of the scientists' increasing reliance upon evidence obtained by laboratory experimentation, contemporary science is often dismissed as being itself just a most basic technology, designed not to deliver natural knowledge but to extend man's manipulative control over his environment, and this is a conception which would deny to man all power of cognition and preclude the possibility of a better understanding of anything at all.

The governing purpose of this book has been to show that the study of evolutionary change began a new era of scientific inquiry. It develops a new sort of natural science, one that makes of every special science the means to a discovery of the creative process that brought man and life and earth and cosmos into being. This discloses the actuality of causal connection between earlier and later occurrence, and returns to us a larger, deeper, and scientifically authentic comprehension of the moral and religious truth which preevolutionary science, in its self-limitation to the observable uniformities of natural occurrence, had to ignore and by implication deny.

X

From Theory to History

Belief in the power of natural science to deliver factual knowledge is today the sole faith to which people everywhere can and must subscribe, and there is no doubt that science has come to stay. Yet this so reasonable creed, which on the one hand steadily extends man's productive exploitation of his natural environment, on the other hand deprives his societies of the cohesive integrity and regenerative power that conditions their persistence, true human association being increasingly displaced by a legal and political authority imposed by physical force. If descriptive science alone can justify its claim to universal and demonstrable truth, must we not look to some major advance made by science itself, one that will give to its verifiable description of fact a new dimension of moral and perhaps religious truth, for the understanding that will restore a dissociate and increasingly demoralized mankind?

To discern its as yet unused potentialities, and discover how descriptive science may be made a source of wisdom and a guide to regenerate life as well as a provider of technical skill, we must somehow reach a new perspective upon science itself, yet it could be objected that descriptive science is no more than a diligent and consistent use of common sense. One might almost say that it arose along with speech, as even the earliest language gave articulate and communicable expression to man's perception of his natural environment. However, the language of science is not just that of everyday speech!

Scientific description uses a language that has a syntax peculiarly its own. In place of ordinary grammar it makes use of theory, and theoretical description had its origin in the rudimentary geometry that gave form and precision to the ceremonial exercises and architectural properties of a celestially oriented religion. The motions of things, no less that the shape of things, can be depicted by means of lines, and even the most primitive geometry could implement a nascent astronomical science descriptive of the heavens.

It was much later, around 300 B.C., that Euclid of Alexandria in Egypt (at that time an extension of Greater Greece) brought the geometrical knowledge of his time into a logically unified system or **theory.** Euclid based his theoretical construction upon a small number of axioms, postulates, and definitions, from which with the help of graphical illustrations he derived the many further propositions comprised within plane and solid geometry. The geometrical theory of Euclid came down the centuries to ourselves virtually unchanged. It was incorporated into mechanical theory and physical theory, and later became the model of the countless theoretical hypotheses devised by modern science. Our forefathers liked to call Euclid the father of geometry, but he might better be commemorated as the father of theoretical science itself.

"Why waste words?" wrote Johann Kepler in 1619. "Geometry existed before the Creation, is co-eternal with the mind of God. Geometry provided God with a model of the Creation and was implanted into man together with God's own likeness."[15] Kepler amplified and confirmed the heliocentric theory of Copernicus, and he has not improperly been called the founder of modern astronomy. Half a century later, in the year 1665 in a pleasant garden at Woolsthorpe, there came the fall of an apple that led Newton to bind into a universal physical theory the postulates of geometry, the principles of mechanics, and his own law of gravitation.

Newton's influence upon modern thought was immediate, enormous, and lasting. Newtonian theory was held to have inaugurated a new Age of Reason, one that would emancipate human intelligence from every traditional belief, and build civilized society upon an innate rational knowledge of the system of natural law that eternally governs and supports all that is.

Modern rationalism was at once more radical and better grounded that had been the classical formalism of Greek and medieval thought, and it aroused criticism more penetrating than any that had gone before. Hume argued that our belief in a causal determination of natural occurrence is only an unwarranted projection into the external world of an ideational process within the mind. Because lightning has usually been followed by thunder, to see lightning is to expect to hear thunder, and we then mistake this purely psychological association of ideas to be evidence of a causal necessity in external reality. Yet we neither know nor can we conceive of any process by which one particular event should literally produce another particular event, and still less can we point to any universal necessity or natural law that determines cosmic occurrence.

Hume's psychology was initially intended to provide a scientific description of mental processes, complementary to that which Newton had given of physical processes, and its skeptical argument is self-defeating, inasmuch as to describe the causal determination of mental events is really to strengthen and not to weaken the belief in its determination of all natural occurrence.

Kant, although awakened as he said by Hume from his dogmatic slumber—by which he meant his rationalistic prepossessions—advanced in his **Critique of Reason** a much more cogent criticism of the basic concepts of Newtonian science, one that would later be in its more important conclusions confirmed and appropriated by descriptive science itself. Unlike Hume, Kant was a productive scientist of considerable ability and some repute before he was turned to criticism by his discovery of certain anomalies or self-contradictions issuing from the fundamental postulates of physical science. Those anomalies, he concluded, preclude any belief in the determination of natural occurrence by some universal geometric or mechanical necessity. What appears to be a geometric structure inherent in external nature is actually a pattern imposed by our own perception and thought upon the chaotic material of immediate sensation. This mental refraction of what impinges upon our senses gives to perceptual experience its spatial dimensions and its definable geometric structure. Further, when inquiry is made into the observable motions of nature, there is added to the three dimensions of space a fourth dimension of time. Space and time are the subjective forms of mental apperception, and the fact that the geometrical and mechanical postulates of descriptive science issue in anomalous conclusions and insuperable problems is proof of their inapplicability to the external world.

The philosophers who ostensibly leaned upon Kant excogitated an idealistic doctrine which identified reality with man's intellectual construction of it. Kant was far too much of a scientist to indulge in such transcendental fantasies, and in his third great **Critique,** that **of Judgment,** he prophetically pointed to another way of scientific cognition, one that descriptive science has now entered but still hesitates to tread.

In the late nineteenth century certain experiments conducted by Michelson and Morley, undertaken to make a more accurate measurement of the velocity of the earth's motion in space, showed that the established concepts of physical science were such as to preclude its exact determination. As a result of those experiments, and of the studies of Lorentz, Larmor, and Riemann which followed from them, Albert Einstein moved in 1915-17 to the formulation of his Theory of Relativity. This radical reconstruction of mathematicophysical theory found in Euclidean geometry and Newtonian physics only special theories not everywhere applicable, only approximately exact, and requiring constant revision or replacement if they are to deliver an accurate description of observable fact. For the experimental physicist in the laboratory as for the astrophysicist intent upon constellations and galaxies, there can be henceforth only theoretical hypotheses which implement increasingly accurate but still only approximate description, and which deliver a probable and no longer an absolute truth. The Age of Reason with its supposed knowledge of the eternal and immutable structure of cosmos is no longer with us, and the true governance of nature, if any such there be, remains to be discerned.

It could be objected that even though no absolute knowledge of physical structure is possible, the continued dependence of science upon one or another descriptive theory points to a more ultimate and basic structure inherent in nature. However, concurrently with the development of physical science described above, there has been a no less revolutionary development of mathematical and symbolic logic which has nullified that objection. This new modern logic little resembles the formal class-logic that came down the millennia from Aristotle. The latter was believed to prescribe the eternal laws of thought, but, in fact, it only provided some useful rules of classification.

As has already been mentioned, the working logic of science has always been theoretical construction. A most significant achievement of the modern logician has been a deeper comprehension of theoretical language. Where Euclid grounded his geometric theory upon a number of axioms, postulates, and definitions from which could then be obtained many further propositions, contemporary logic somewhat similarly finds a theory to have its nucleus in a small set of primitive sentences, from which can be constructed an unlimited number of other sentences making use of the terms contained within the primitive sentences. A theory is best understood as a large and complex definition of terms, exemplifying how these terms may be used in meaningful statement. The function of theory in descriptive science is to provide a means of symbolic expression, i.e., an appropriate language. However, a perfectly good theory may define terms which have no current or conceivable application in factual description. Ordinary grammar comprises a rudimentary logic, and, much as a literary fantasy may be perfectly grammatical yet describe a realm like nothing in human experience, so may a perfectly good logical theory have no current or conceivable use in scientific description. The logical structure of language, whether this be grammatical or theoretical, is a character of symbolic expression, and it is a mental construct that need reflect nothing and have no correlate in external nature.

In this way modern logic confirms the conclusion reached by contemporary science. Where the latter shows the notion of an overall cosmic structure to be illusory, logic and linguistics show how this error arose and why it so long retained its hold. Belief in the particular determination of later by earlier occurrence remains to support and direct scientific investigation, but this leaves no place for any general or universal determination of natural occurrence.

These thoroughgoing changes, which might have been held to remove the solid foundation upon which had been firmly established all earlier thought, did not lessen but rather strengthened faith in descriptive science, for they showed that this scientific faith springs not from the supposedly self-evident axioms of geometric theory, but from the power of science to proceed from observable fact to verifiable prediction. In place of the illusory belief in some universal law or cosmic necessity that shapes all that is, there comes a well-founded faith in the particular necessitation of later by earlier occurrence, and this is all that science needs.

There does remain one discipline, that of arithmetic, to which can and must be accorded an absolute authority. This exceptional status is due to the fact that arithmetic takes account only of the ubiquitous differentiation of nature, and postulates no real or imagined uniformity of natural occurrence. The particular differentiation of nature is what makes things discernible, comparable, countable and measurable, in short, knowable. It gives to nature its ineradicable manyness or plurality. Whether arithmetic should be accounted a part of descriptive science is debatable, for what science would describe is cosmos and what arithmetic describes is the ineradicable manyness or plurality of nature, which is to say chaos, i.e., a realm that has no likenesses or uniformities of character. Arithmetic should remind us that the chaotic differentiation of nature is also an indispensable condition of its knowability.

Nevertheless, although it may not weaken scientific faith, the recent shift from Newtonian to Einsteinian science does undercut the expectation that theoretical science will ultimately deliver a consummatory knowledge providing comprehension of our world. There is no longer any possibility of a single all-comprehensive theory of nature. Yet it was this expectation that had given to science its impetus and sustained momentum, as well as justifying its high evaluation as an activity to be pursued for its own sake. With that expectation gone, will not descriptive science deteriorate even more rapidly into a mere technology that seeks only control and not understanding of our world?

The alternative to this probably disastrous outcome is to open some new approach to natural knowledge, one that may make use of our growing theoretical knowledge of natural kinds and local uniformities to reach an understanding of how and why these general characters should have arisen and persisted. This science would provide a knowledge of the creative processes of nature, it would be a study of creation itself, and it is one that has been pursued for more than a century under another name. It is the so-called evolutionary science inaugurated by Darwin's publication of **The Origin of Species** in 1859; and it already tells us a good deal about the creative process that brought into being the vegetable and animal life on our planet.

It is regrettable that the symbolic logician has not yet thought to analyse the sort of language that makes possible historical description. Such a study would make it clear that historiography, although it makes use of whatever theoretical knowledge may serve its descriptive purpose, is advanced not by theoretical hypothesis but by historical hypothesis. An historian has no peculiar gift of historical intuition, he disposes over no esoteric wisdom or special mode of inquiry. He, too, must depend upon the scientific method which draws hypotheses from and confirms them by means of observable particular fact. His distinction comes from his subject-matter, which requires its own sort of symbolic description, for history must describe the causal processes by which collective units such as species, living populations, and societies are originated, undergo change, and either perpetuate themselves or become extinct. Up to the present, this historical study of the creative processes of nature has limited itself to the description of vegetable and animal history, but to extend it to nature at large will be to seek some comprehension of how cosmos was created, and this is a study which has no stop. It will provide true understanding as well as knowledge, and it may indeed make of man a second Adam who has learned to divide good from evil.

XI

The Evolution of Inorganic Nature

The preceding chapter stressed the need of a further development of evolutionary science, extending its present description of plant and animal history to cover the long evolution of material nature. Just as the observable uniformities of living organisms have made possible an historical description of the evolution of species, so may the many uniformities found in wider nature support a description of the creative evolution that brought a cosmos into being.

There is a good reason why evolutionary science should have been first developed by life-scientists, and still be generally regarded as no more than an outpost of life science. It is simply that living organisms are large enough to be visible to the naked eye, so that their individual differences [or "variations"] have always been apparent along with the similarities that bring them into their natural kinds, this then allowing the construction of a classification of species. In this way, attention was finally directed to the causal relationships that bind individuals into natural groups such as species, and that make of these differentiated individuals the effective determinants of evolutionary change (which, properly speaking, is just group change).

Although the microscope and other more sophisticated aids to perception have importantly deepened the range of human vision, no instrument as yet makes visible the individual differences of physical and chemical units such as light waves, protons and electrons, atoms, and molecules, and there has consequently been no inquiry into the causal processes that bind these natural but indiscernible units into the material groups that give to inorganic matter its seemingly indefinite persistence.

Accordingly, although there is increasing and even convincing evidence of evolutionary change occurring in nature at large, there has been no sustained effort to describe it. Preventing this was a long-established tradition, transmitted from Greek antiquity, that required matter to be indestructible, and to be composed of immutable atoms that undergo no change other than that of their spatial position. Greek atomism had adequately supported the initial development of modern chemistry, but it had to be relinquished when the discovery of radioactive elements showed that certain atoms may be transmuted into other atoms. It was also established that the sun owes its luminosity to a continual transmutation of lighter hydrogen atoms into heavier helium atoms, again with the emission of a residual energy in radiation.

These and other important advances of physical science should have stimulated more searching inquiry into the nature of chemical matter, for they give quite new significance to what had earlier been learned of it. The table of elements brings more than a hundred elements into a linear series showing their increasing atomic weights. The lightest element is hydrogen, and the heaviest, uranium, has more than 200 times its atomic weight. This points to a long evolutionary progress that made of smaller atoms a material that was repeatedly given new organization in larger and more complex atoms, and this material progress resembles that of living nature, which out of microscopic bacteria and protozoa produced multicellular organisms which came to be steadily larger and of more complex organization. Does not this provide adequate evidence of a continuous evolution that brought first matter and then life into being? Must we not infer that the causal processes which brought about the evolution of life only modified those which had earlier advanced the material evolution that led up to and into life?

Moreover, as Mendeleef first observed, when the elements are so arranged this serial order makes apparent a repeated pattern, due to the fact that atoms of widely disparate size and weight fall into periodic groups which display a remarkable similarity of chemical behavior. Mendeleef attributed this to a natural law of periodicity universally imposed upon material nature, but to an evolutionary scientist it points rather to a genealogy of atomic kinds, or alternatively to a convergence of these different kinds to very similar types of external adaptation and visible behavior.

Physical theory contains within its most basic postulates two principles which appear to stand in verbal contradiction. The first of these requires an absolute conservation of energy throughout all physical change, and stipulates that the quantity of energy issuing from any change must be exactly that which went into it. This postulate allows a description of natural change by means of quantitative equations. The second principle, which reformulates an earlier law of thermodynamics, is that of entropic increase. This states that the amount of usable energy, i.e., energy able to do work and effect further change, is steadily depleted, what issues from any physical change being less by a determinate amount than what went into it. But energy is defined as the power to do work, and we must infer that available energy, notwithstanding its absolute conservation, is steadily used up even by its use, part of it always being converted into something other than energy.

The physicist does not fail to draw the large implications of the law of entropy. In its universal application, he tells us, physical theory must describe a world in continuous dissolution. Energy is everywhere expended, larger and more complexly organized material units being steadily disintegrated into smaller units, and a time must come when no effective energy remains available, and there can no longer occur any describable change. We are told of a primeval cosmos that was fully organized and replete with energy, but that steadily dissolves into an indescribable immaterial chaos.

The principle of entropic increase thus discloses itself to be in fact an historical hypothesis, one that requires the scientist to describe a universal process of devolutionary decline and extinction. This does not preclude the possibility of there being local progresses here or there, but it does close every prospect of indefinitely sustained progress. Time, too, must eventually have its stop.

Much as the principle of absolute conservation looks like a linguistic convention allowing a descriptive use of quantitative equations, so this hypothesis of cosmic devolution may well turn out to be descriptive of scientific method rather than of natural occurrence. The physicist is as yet wholly dependent for his description of external nature upon an initial evidence obtained by means of laboratory experimentation. The latter entails a tearing of inorganic matter from its natural context and its removal to the laboratory, where it is subjected to a destructive analysis that breaks down its larger units into their component parts—crystals and colloids into molecules, these into atoms, these into protons, electrons, mesons, and these finally into waves of radiant energy. Is it not to be expected that this experimental evidence must primarily disclose only a devolutionary process, one that is actually that of the experimental procedure itself? And is there to be found among the postulates of physical theory any hypothesis that would allow a description of evolutionary progress, as do those which have supported the description of the evolution of planetary life?

Most basic to this latter description are the postulates that define evolutionary change as a process that occurs in living groups as a consequence of observable differences of character found among their individual constituents. In order to describe an evolution of material nature, we would need to discern the natural groups that give to inorganic matter its many uniformities of character and its enduring persistence, and to do this we must reach a new conception of the individual units that compose matter.

We are aware today that radiant energy consists of waves, these being individual units each of which produces a successor to which it transmits its determinate length. Light moves almost incredibly fast, but it does take time, and each new wave needs a moment of time in which to complete its length, which is to say its growth, with longer waves taking more time so that different sorts of radiation may still travel at the same speed. As we also know that a disintegrating atom emits radiation, we must infer that chemical atoms were originated by means of a reorganization of radiant energy, the units of which were given a circular or elliptical motion such as is now ascribed to electrons, in order to make of electrons and other units the component parts of an organized atom.

Having reached this conclusion, we should not still suppose these material units, compounded as they are of radiant energies, to be indefinitely long-lived. The larger plants and animals are generally longer-lived than are the smaller, and take more time to grow to maturity before they are reproduced in progeny, and we should infer that matter everywhere, in its so-called inorganic no less than in its organic domains, owes its persistence to a continual reproduction of its short-lived units, the temporal dimensions of these being on the whole proportionate to their spatial dimensions or size.

But we should also improve our scientific terminology, now that we have replaced the old misconception of solid and immutable atoms by a conception that allows us to recognize the highly organized character of all chemical units. No longer should we perpetuate the old dichotomy of organic and inorganic nature, which necessitated the supposition that life must have come to this planet on a meteorite driven in from outer space. There has been a continuous progress here on or in earth from the smaller to the larger units of chemical matter, and eventually to the still larger units of biotic matter. We are hereby informed of a material nature that lives throughout its length and breadth, and that has always owed its persistence to the reproduction of individual units comprised within natural groups which undergo evolutionary change.

Our visual perception of matter must resemble that of a visitor from outer space who looks down upon Amazonian forests. Below him there stretches to the horizon a vast jade floor, broken perhaps momentarily by the glistening of a river-bend in the sun, and, if this is all he knows of earth how should he guess at the sprouting seed, the fallen trunk, or the manner in which this dense vegetation feeds upon the loam that is its own decay? Somewhat similarly, our visual perception of a solid and smooth sheet of steel has had to be replaced by the mental image of a diffuse cloud or open lattice, through which a stream of particles may pass virtually undisturbed. We must now further exercise our powers of imagination, in order to conceive of this open lattice-work as a pattern of spatial distribution maintained by short-lived individuals throughout their continual reproductive change.

We know in nature nothing immutable, nothing immobile. Any material object we may closely observe compounds into its particular motion the motion of the earth's surface, that of the earth in its orbit, that of the sun in its galaxy, and that of the galaxy in space, and any apparent constancy of form testifies to the fidelity with which short-lived and growing individuals may transmit their own character through successive generations. We should not suppose the theory of relativity to establish the relativity of motion, as it only makes explicit the inevitable relativity of our own measures of motion. Motion itself is real enough, and it may be the ultimate substance of things, brought to a formal and describable nature from the chaotic change that still persists in the ineradicable particularity of what is.

The astronomer Fred Hoyle and his associates, in an

endeavor to account for the relative absence of all heavier elements in the older stars, have advanced a theory of "steady state" creation which would offset and complement the prevailing conception of a universal dissolution of material nature.[16] This theory postulates a continual and spontaneous production of hydrogen atoms within the widening intergalactic spaces of an expanding cosmos. Gravitational attraction exerted by the galaxies as they sweep through space gathers up wide swathes of this newly created and uniformly distributed elemental matter, and this accretion makes good the steady loss of available energy due to entropic increase. No description has been reached, however, of a creative process that might produce those new hydrogen atoms out of immaterial space.

Another indication of the steady approach of physical science to a theory descriptive of evolutionary change may be found in its increasing use of statistical description. This method was developed by eighteenth century thinkers as a means to the graphical depiction of rising birth rates, mortality rates, the incidence of epidemic diseases, the changing consumption and production of commodities, and other social trends. It was later to become implemental to scientific description in many fields, and it promises to provide a language appropriate to all description of natural occurrence. It may fairly be said to have created the sciences of meteorology and economics. The paleontologist has used it to make of fossil evidence the means to a graphical depiction of the expansion and decline of plant and animal groups. The astrophysicist relies upon it for his description of galactic systems, and during the last few decades it has implemented the geneticists' description of evolutionary change in plant and animal species.

Although he is not able to identify or delineate the natural groups that perpetuate chemical matter, the microphysicist can nevertheless make use of statistical description by means of a methodological concept, that, namely, of "populations of probabilities." Inasmuch as probabilities are not individual units, but only fractionally expressed chances of possible occurrence, which is to say they are mental constructs and not concrete actualities, that concept is actually a useful scientific fiction. It nevertheless can and does implement verifiable prediction, and it points the way to a scientific description of the natural groups which compose material nature and everywhere undergo evolutionary change.

To conclude this brief indication of a few of the ways in which theoretical science already anticipates an historical science descriptive of cosmic evolution, mention may be made of a striking analogy that likens the later evolution of chemical matter to the early evolution of plant and animal life. The progress of chemical matter was advanced by way of progressively larger units, atoms being followed by micromolecules, macromolecules, and finally crystals and colloids.

The indefinitely many molecules organized into a crystal exert reciprocal forces which keep them at safe distances in their serried ranks, and this is a mode of organization that supports growth to very considerable size, so that crystals may far outrange the microscopic bacteria and protozoa that led up to plant and animal life. Nevertheless, crystalline matter testifies to a wide divergence from the line of evolutionary progress that was to proceed through the colloids to the plants and animals. The solid-state physicist learns that physical units are always organized into the composite patterns that best preserve them, and the intermolecular forces which give to the crystal its internal rigidity also work within the molten fluid, although there they are weakened by heat-vibrations which give to the molecules a limited mobility (as is observable, for example, in water, which visibly reveals its crystalline character only when it freezes at low temperatures).

In the colloids the advance to larger unit-size was made at less expense of molecular mobility. We are familiar with colloids in muds and slimes, in oils, creams, fats, and jellies, and in all bacterial, protozoic, vegetable, and animal life. The organization of the colloid is such as to keep this still quasifluid or amorphous, so that it remains highly plastic, and it was just that plasticity, of course, that made possible the long progress through the motile animal organism to human life. The divergence of the crystals from the line of colloidal progress is an earlier instance of the sort of divergence that would later divide vegetable from animal life, for this divergence was likewise due to a sacrifice of the power of locomotion in exchange for the adaptive habit that roots the plant to its nourishment in the soil.

Today, there is cumulative evidence testifying to a continuous and describable progress that was advanced within and by the materials composing the surface of the earth, and that led through atomic, molecular, and colloidal units to the lower animals, the higher animals, and our human selves. We may now turn to a consideration of some of the deeper implications of this planetary history, for these are such as to bring to us a new and better comprehension of human nature, and to remind us of the creative powers and responsibilities that came to man as a heritage from his long past.

XII

Man in His Environment

Today we are all ecologists, increasingly concerned, as we should be, by the harm done by city wastes and industrial poisons to an environment that must still provide our every sustenance. Man's immediate environment is the surface of the earth with its continents, oceans, and atmosphere, but his whole environment can be measured only by his effective knowledge, which reaches to the stars and further extends a steadily empowered technology.

Just how are we to conceive of this cosmic environment? Descriptive science, although it tells us much about the spatial distribution of cosmic matter, says little about the temporal dimension that is its history. Great telescopes find no end to cosmos, but they do allow a calculation of the distances that separate other galaxies from our own, and the spectroscope, that remarkable instrument, informs us that the distance of any galaxy is also a measure of the velocity that carries it still farther away.

This has given rise to the hypothesis of an expanding universe, one which states that the galaxies become ever more widely dispersed, somewhat as do the spots on a balloon that we inflate. On the other hand, Einsteinian physics at present requires the hypothesis of a closed cosmos. What were earlier taken to be the infinite media of space and time have shrunk into the variables of a scientific formula, so that they now describe only particular and local change. These two very different cosmological conceptions are not necessarily incompatible. The seemingly rectilinear motions of the galaxies may yet turn out to be circular or elliptical, so that even the most distant galaxy will eventually turn back home, or it could be that the fastest moving galaxies are actually approaching their time and place of extinction, beyond which is nothing describable.

Spectroscopic analysis delivers other important cosmological information. It shows that the composition of cosmic matter is everywhere much the same as that of earth, as it is constituted of the same sorts of atoms, and it also discovers a notable absence of heavier atoms in the older stars. This latter is a fact of great significance, as it directs us to a new and rather startling description of cosmic history. One might have expected to find a dearth of heavier matter in the younger stars, on the reasonable supposition that these have not yet had time to complete their chemical evolution, but just the opposite is the case, and we are thereby led to a very different conception of the evolution of cosmos.

There evidently somewhere proceeded a sustained progress from lighter to heavier matter, with smaller atomic units repeatedly reorganizing themselves into larger, heavier, and more complex material units. This continuous and directed progress was necessarily the achievement of a single natural group, that namely which would later proceed through further evolutionary change and create the molecular, crystalline and colloidal, and eventually biotic units of a still material world, for the growth of even the largest plant or animal is still conditioned by the behavioral activity and character of its constituent atoms. In short, we must conceive of a single creative progress which was advanced by the continuous evolutionary change of just one natural group, and the pattern of cosmic history accordingly must resemble that with which we are now familiar in the evolution of species.

In this terrestrial history, we have seen, there proceeded an evolution of plant and animal kinds that testified to a sustained progress advanced by the singular group that was eventually to be widely dispersed in mankind. From that creative group there repeatedly fell away marginal populations which, having diverged from the direction of evolutionary progress, were then expansively multiplied into great categories of more or less closely related species. We like to say that history repeats itself. Literally taken, this is quite inconceivable, but it can be understood to mean that later history may follow a course strikingly similar to that of earlier history. In this sense, the history of plant and animal life may be said to repeat that of the chemical evolution which brought out of smaller atoms the larger and heavier atoms.

In the earlier evolution, that of chemical nature, there repeatedly diverged from the direction of evolutionary progress certain atomic populations which, as a result of their separation from their creative progenitor, were expansively multiplied into enormous aggregates, and it was these great clouds of atomic matter, later condensed by gravitational attraction, that became the stars. Because the stars originated in this manner at earlier and later times, they bear witness to successive stages of the chemical progress continuously advanced by their common progenitor, and this is why the older and earlier originated stars testify only to the earlier progress, while the younger and later originated stars testify also to the later progress, and contain both light and heavy elements.

This is an historical retrospect upon the evolution of chemical matter which lends little support to the popular scientific fiction now current. This tells us of civilizations supposedly existing on the planets of suns so remote that they must be approached by means of spaceships travelling at velocities close to that of light, so that time itself is slowed down and the astronaut may return to the earth some decades or centuries after he left it. If our retrospect upon chemical history is anything like the truth, those tales are but the most errant fantasy, for there could have been no divergence of terrestrial populations to outer space later than two billion years ago, when biotic matter came into being and earth was consolidated very much as it is today.

However, there are also astronomers who subscribe to this belief that life somewhat like our own must exist elsewhere than on earth. Given myriads of galaxies each containing billions of stars, they reason, must there not be innumerable suns much like our own? Must not many of these suns have planets, and some of these planets closely resemble earth? Is there not a quite overwhelming probability that cosmic matter everywhere sporadically blossoms into colloidal, sentient, and finally intelligent matter that inquires into how it and other matter came to be?

But this reasoning is wholly specious, inasmuch as a single instance of anything establishes no probability whatsoever, whether this be great or small. We know of but one sun, one solar system, one planet that supports plant and animal life. There is just no evidence at all to support the hypothesis that such life probably exists elsewhere than upon this earth. There is, to the contrary, as yet only overwhelming evidence against that supposition, for what is found outside of earth is always and only a process of devolutionary change, this being a process that disintegrates all evolved and organized matter into its component parts, and this is a change that must eventually return cosmos to chaos. All the evidence we have of natural occurrence supports the hypothesis that there was never but one sustained direction of creative progress, that this progress was directed toward and issued in human progress, that the progress was advanced by a single group contained within the substance of this planet, that it was always achieved by means of larger and more complex organization, and that it is advanced today by man's faith in his human destiny.

The sun and its planets occupy an inconsiderable and inconspicuous place near the periphery of a rotating galaxy, and they cannot very well be called the spatial center of anything. But the steering device of an automobile or oceanic liner or transcontinental aeroplane is not centrally placed, nor as a rule is the governing capital of an empire, nor is the brain of man. The living group that was ancestral to all protozoic and animal life could probably have been contained within a cup of water, and it could not have spatially centered anything. A purely theoretical science must conclude that the large uniformities of nature establish and govern the smaller, but an historical science such as describes evolutionary change finds just the contrary to be true.

We should be more appreciative than we are of this earth that is our human habitat. The more the geologist learns of earth, the less he finds it to be representative of other nature. There is nothing in the uniformities of nature or in the general properties of matter to require that there be a planet of just this optimum size, heavy enough to retain a light atmosphere but not so heavy as to generate destructive gravitational pressures, close enough to the sun to receive beneficent warmth but far enough away to escape excessive heat, maintaining a speed of rotation so braked by the moon that it secures an optimum benefit from the dispersion of solar heat by wind and water currents, possessed of a light and porous crust safely confining its hot and heavy core, given mountains that vent excessive heat and an atmosphere ideally adapted to transport light and warmth but to smother harmful radiation, and also working to prevent a too close approach of the moon that would otherwise bring catastrophe to earth. These many varied conditions do not only support plant and animal growth, they have equally conditioned the composition and maintenance of the mineral surface of the earth.

Most consequential among these geological adaptations that have preserved the earth and supported its long evolution was the retention of a goodly amount of its first originated and lightest element hydrogen. At all but the lowest temperatures hydrogen is a gas, and the rapid motion of its light molecules would ordinarily make it the first element to be rapidly dissipated into outer space. This was prevented by its molecular combination with heavier atoms, so that it made possible the film of water that would come to cover most of the earth's surface.

The role played by water in the evolution of the earth

seems out of all proportion to its small contribution to the earth's mass, and the later evolution might reasonably be depicted as a continuation of aqueous history. Although the oceans may seem enormous to ourselves, a geologist has likened them to the dew left upon a large pumpkin, but, as L. J. Henderson was at pains to point out two or more generations ago, the properties of water are in every respect unique.[17] They make of water the most efficient stabilizer of temperature known or conceivable. It was oceanic and atmospheric water that made and kept the earth's mineral surface porous and light, so that it did not become congested and generate destructive heat. Water alone made possible the aqueous colloids that paved the way to plant and animal life, of which water remains the major constituent, most of this biotic matter being three parts water.

But perhaps the most impressive achievement of water was its creation of the great thermostatic system that preserves the earth. By means of this self-regulating hydraulic apparatus, the poles and tropics, oceans and ocean-currents, atmosphere and wind movements, sunlight and evaporation, rain and snow, mountains and forests, rivers and drainage systems all cooperate to one end, this being the prevention of those harsh extremes and rapid shifts of temperature that would otherwise disintegrate terrestrial matter. Through the workings of this vast hydraulic system local warmth is quickly dissipated, surplus heat is channeled off, seasonal and diurnal temperatures are moderated and graduated, and the susceptible surface of this living and sensitive earth is effectively insulated from the destructive heat that is generated within its interior and that impinges upon it from without.

This geologic ecology was a good many billion years in the making and it has constituted a large condition of the creative evolution that fashioned earth, maintaining throughout planetary history its effective service. This history can be described, to allow a discernment of the sorts of causal process that advanced it, but this does not mean that creative change can be wholly accounted for. Its description must still be couched in general terms which communicate a knowledge of the uniformities of natural occurrence, and we have learned that forms originate in individual variations, which is to say in the particular differences of members comprised within natural groups. These particular characters, unlike the general characters or forms of things, remain quite unpredictable, and, if we may define a miracle as an event that could not have been predicted and which must remain in some degree inexplicable, we realize that the creative evolution of cosmos presents a continuous and ever-present miracle. It is something to be appreciated and properly valued, rather than explained away.

Life science will probably always provide our most authentic description of the causal processes and local occurrence of evolutionary change, and within its special domain, which might be called the genic realm of planetary life, evolutionary change is found to result from a continual diversification of individual character due to sexual reproduction and gene exchange, this ensuring a random selection of just one particular recombination of parental genes out of a truly incalculable number of possible recombinations. In relation to this a biologist writes: "Assume that at the end of every second during the last billion years the entire existing flora and fauna has been liquidated and replaced—would we then have exhausted nature's possibilities? Alas no. In spite of this valiant effort to be extreme, we have scarcely nicked the surface of the possible."[18] (Why "alas," if this genetic interchange is what advanced the creation of planetary life?) We are aware that to assess the probable, we need exhaustive knowledge of the possible, and we are thus informed that no knowledge of earlier occurrence would at any time have made predictable the further course of natural creation.

This applies no less to the present time than it does to any past time. However, although our knowledge of the past course of creative change leaves the future unpredictable, it may nonetheless well instruct us as to what we would have that future be, and what we should seek to make it be. Historical retrospect does deliver a new understanding of man's place and function in nature. It shows his location, which lies within a natural group the continuous history of which goes back to the beginning of life and presumably to that of measurable time. But this might in some sense be said of every existing group, and the particular group that had issue in mankind must be distinguished as that which had advanced the sustained progress toward human life. This group history is unique, every other history being divergent from the direction maintained by its creative progress. The group ancestral to man is the group ancestral to all other natural populations living or extinct, and its function has always been that of material creation. It was always the responsible tenant of a creative power exercised by, within, and from itself, and so, surely, should it remain. So would we have it be!

Are we of this epoch not fortunate, living as we do at a time when science has come of age, and advances from a purely theoretical knowledge of the uniformities of nature, this allowing at best a probable prediction of future occurrence, to an historical science truly descriptive of the course of creation? With science, man himself has come of age! It is descriptive science, supported by the entirety of factual evidence, that requires his acceptance of an inherited responsibility for the continuance of creation. Must he not find his creation good, and thereby become a new Adam everywhere able to divide good from evil? This is a responsibility that comes to each human individual, to every human society, and to the society of human kind that may emerge today. Its acceptance will bring to new expression many earlier insights of religious faith and cultural tradition, and incorporate these into a single human faith that can give to humanity a new and needed goverance. Truly were we told: The kingdom of heaven lies within you.

It is surely providential that today, at last, man is compelled by self-interest to look to his environment, properly assess the havoc he has worked upon it, and seek and find how his house and habitat may be set in order. Our most pressing human assignment is to preserve this earth and as much as we can of its living tapestry, for what persists on earth must henceforth be just what man cultivates or protects. While pursuing this task, we may also hope to reach a knowledge of the place and function of our planet in the larger economy of nature. How far may not this little candle that is the earth cast its redemptive beam into a dissolute and naughty world? We know as yet nothing of the interstellar forces that work within a galaxy, to hold matter so widely distributed that it does not congest and quite rapidly annihilate itself.

We do know how the smallest causes may trigger off the largest effects. It takes only the smallest trace of water-vapor to bring to spectacular explosion the holdful of gas remaining in an empty tanker at dock, and a microscopic virus can let loose an epidemic that decimates a continent. Slower working and less spectacular, but immeasurably more effective, have been those individual variations of group character which, since time and creation began, have determined the routes of evolutionary change, to make of earth and cosmos what they are today.

The actuality revealed to us by an historically oriented science of evolution is in its most important features the diametrical opposite of what earlier science reported it to be. Is it inconceivable that this earth should yet give purchase to a lever that will turn the stars in their courses, and reverse the process of devolutionary change that would return all that is to chaos?

But, to begin with, it is ourselves we must redeem. Our immediate need is of an intellectual revolution that will replace long established misconceptions by a truer description of nature with a proper appreciation of man's creative responsibility within it; and with this intellectual advance must go a behavioral progress that will make of a divided and self-destructive mankind a fully communicant and truly associate humanity. This may call for new symbolism, to be provided by a creative art that will make of domestic nurture within the human family the initial phase of a responsible and wise education, one which recovers and incorporates into its teaching every cultural achievement and religious insight that has given to man a faith in human integrity and a prophetic awareness of his cosmic importance. And all of this may require the supervention of political and militant power by a supreme authority able to control and direct a human economy that appears about to destroy itself and earth alike. All this and more needs to be done, but nothing that a human providence may not still provide. Yet let it be done soon and quickly, lest there fall that night in which no work is done!

Notes

1. For the development of population genetics see **Genetics and the Origin of Species** by Theodosius Dobzhansky (New York: Columbia University Press, 1937, 1941, 1951); the pioneer study by Sewall Wright in **Genetics XVI,** pp. 97-159; **The Strategy of the Genes** by C. H. Waddington, 1957; **Behavior Genetics** by John L. Fuller and W. Robert Thompson (New York: John Wiley & Sons, 1960); **Variation and Evolution in Plants** by G. L. Stebbins (New York: Columbia University Press, 1950); **Evolution After Darwin,** Vols. I, II, and III, edited by Sol Tax (Chicago: The University of Chicago Press, 1960); and the many issues of the journal **Evolution,** published by the American Museum of Natural History, New York.

2. For an excellent study of the fossil evidence with graphical illustrations of the group histories see **The Meaning of Evolution** by George G. Simpson (New Haven: Yale University Press, 1949).

3. **The Vertebrate Story** (revised edition of **Man and the Vertebrates)** by A. S. Romer (Chicago: The University of Chicago Press, 1959).

4. L. W. Evans in **Group Processes,** edited by B. Schaffner, 1955.

5. **A Study of Instinct** by Nikalaas Tinbergen (Folcroft, Pa.: Folcroft Publications, 1950).

6. See the article by H. W. Nissen in **The Non-Human Primates and Human Evolution,** edited by J. A. Gavan, 1955.

7. Tinbergen, **A Study of Instinct.**

8. Charles Darwin, **The Origin of Species and the Descent of Man** (New York: Modern Library), p. 435.

9. Konrad Lorenz in **Group Processes,** pp. 208 ff.

10. C. H. Waddington in **Evolution After Darwin,** vol. II, p. 148.

11. A. I. Hallowell in **Evolution After Darwin,** vol. II, p. 333.

12. The incantation is taken from **Man Makes Himself** by V. Gordon Childe (New York: New American Library, 1952).

13. G. R. Willey in **Evolution After Darwin,** vol. II, pp. 133-36.

14. Arnold Toynbee, **A Study of History** (Oxford University Press.)

15. Johannes Kepler in **Harmonice Mundi** as quoted by Arthur Koestler in **The Watershed** (Garden City, N.Y.: Doubleday & Co., Inc., 1960).

16. **Frontiers of Astronomy** by Fred Hoyle (New York: Harper & Row, Publishers, 1955).

17. The **Fitness of the Environment** by Lawrence J. Henderson (Magnolia, Mass.: Peter Smith Publisher, Inc.).

18. **Nature and Man's Fate** by Garrett Hardin (New York: New American Library), p. 287.